Fixing Einstein's E=mc-squared

Replacing Observed Mass ('m') with the 'M' Nucleus Magnetic Force Divided by the Volume of the Electron Shell Radius Separation

At $d = R_{ES}, v = 0, t = 0$ Kelvin, Weak Force Equilibrium

$$\frac{kQ_zQ_e}{(R_{ES})^2} + \frac{M^2(z_1, neu_1)(z_2, neu_2)}{(R_{ES})^3} + i\frac{M^2(z_1, neu_1)(z_2, neu_2)}{(R_{ES})^3} = 0$$

$$M(z_1, neu_1) = Q_1\sqrt{kR_{ES}}$$

$$Gravity\ Force = \oiint_t \left[\frac{kQ_1Q_2}{(d)^3} - \frac{kQ_1Q_2}{(d - R_{ES})^3}\right]$$

$$mass = \frac{M^2(z_1, neu_1)}{2 * \frac{4}{3}\pi(R_{ES})^3} = \frac{Q_1 * \sqrt{kR_{ES}}}{2 * \frac{4}{3}\pi(R_{ES})^2}$$

$$Free\ Energy = c^2 * \frac{M(z_1, neu_1)}{\frac{8}{3}\pi(R_{ES})^3} = c^2 * \frac{Q_1\sqrt{kR_{ES}}}{\frac{8}{3}\pi(R_{ES})^3}$$

By Arno Vigen

© 2017/10, 2018/04 Arno Vigen

Simple Words to Understand . . . Chemistry, Elements, and Bonds

Why does a Nucleus Stay Together If Protons Repel?

- A Nucleus is Just . . . a Magnetic Ring

Why Don't Electrons Fall into the Opposite-Charged Nucleus?

- Electrons are Just . . . Frightened by Nucleus Magnetics

Electron Shell Chemistry Is Just . . . Scrunched Cube Geometry

- Why are electron shells in sets of 2, then 8, then 8 and such? Can we improve Pauli-aufbau?

Scrunched Cube Electron Shell and Bonding Periodic Chart of Elements

- Understanding Molecular Bonding in the Scrunched Cube Atomic Model

Scrunched Cube Molecular Bonding

- Understanding Molecular Bonding in the Scrunched Cube Atomic Model

Scrunched Cube Explains in 3D What Makes Molecules Solid, Liquid, or Gas

- And Why is Gas of Every Element the Same Volume (a mole)?

Simple Words to Understand . . . Gravity and Other Forces

Gravity is Just . . . That Electrons are a Little Closer

- Explaining Gravity from the basics of Electromagnetism and Explaining Why Observed Mass Changes

The Five Continuous Fundamental Electromagnetic Forces: Reconnecting Newton into the Chemistry and Particle Physics

- Resolving Strong Force, Weak Force, Bonding Force, Gravity, Mass and $E = mc^2$ via the basics of Electromagnetism as One Continuous Function

Fixing Einstein's E=mc-Squared

- Integrating that Mass is Magnetics divided by the Electron Shell Radius Volume Explains

Solving Schrodinger's Equation with Nucleomagnetics

- Introducing a change to the electron distribution based upon the AVSC nucleomagnetics postulate

How is Electricity and Magnetism Linked?

- Exploring the Fundamental Linkage of Charge and Magnetism

Does Time and Space Really Warp?

- Replacing Electron-Shell Radius for Time-Space Factors in formulas such as the General Theory of Relativity

Simple Words to Understand . . . Personality

Visual Astrology: Fun, Support, Security, and Growth

- Astrology 'signs' archetypes are based upon powerful traits to understand people

Visual Astrology Relationships

- What happens when 'sign' personalities interact

Visual Astrology and Jung

- Astrology 'signs' archetypes actually predict all the Jungian 4 archetypes

Dominant Personality Traits

- Dominant Personality Traits Follow from Four Dimensions, Six Steps and so 24 Subcategories

Simple Words to Understand . . . Communications

Decision Matrix® Writing

- Persuasion is based making arguments and comparisons at the correct strength in an understandable, powerful order.

GATESOUP® Writing

- **G**oal, **A**udience, **T**heme, **E**nough **E**lements, **S**upport and the rest

Kedarf® Grammar and Composition Explained

- Defining the Parts of Speech, Paragraph Structure and More in Usable Terms

Table of Contents

Challenge: For Newton's and Einstein's Formulas to Work Well, 'Mass' Cannot Keep Changing!...9

    Observed Mass Works – That is not the Problem 9

    Not that Complex..10

    'Mass' is a Shortcut 11

    Converting Total Magnetic Force of Particles to Mass 14

    AVSC Two-Part Mass is Quantized by the Number of Particles.......15

    Electrostatic Charge and Nucleomagnetic Forces are Not Infinite..15

    Current Cause of Mass Not Known..15

Hypothesis: All Energy comes from Electromagnetics – Electrostatic Charge and its AVSC discovered Newton Complement, Nucleomagnetics...21

Relating Electrostatic and Nucleomagnetics Force............................33

    RES Changes in Direct Relationship to Nucleus Particle Count.......35

    Because the direct relationship goes with nucleomagnetics, observed mass prefers that measure 35

    Because the overpowering force at a distance is electrostatic charge, the gravity calculation uses 1/distance squared.................36

    Notice that this is Average Radius of the Electrons Shell – not Covalent Bonding Distance, not Edge of Molecule, not Outer Shell 37

    Root-G and Root-k Special Uses..39

Calculation of Gravity from Charge versus Newton 6.674×10^{-11} $m3kg(s2)$ 42

If the Universe is in Balance, How Can There Be Overall Net Attraction?..52

Fundamental Question: Why does 'mass' change at speeds near the speed of life?...55

Fundamental Question: What Explains why Mass Changes in Bonding? 57

What changes is R_{ES}, not (ZN) (the number of nucleus particles). 57

Bonded Electrons do not contribute to the Full External Magnetics (Atom-gravity) 'mass' observed, Instead Push Nucleus Apart 59

Decrease offset in that Electron still contributes to push out other electrons 60

Must also realize location is not fixed, but a quantum path 63

What is the Einstein's Energy? How do the Units of Measure Work! 64

Redefining the Quantum of Energy 68

Counter Argument: At zero, the nucleomagnetics become infinitely larger, so how can they balance. 69

Calculation of Quantum of Energy is the Adding of One Electron to the System 74

Reflection of New Particle Creates Other Particle Changes, So Field Change Electromagnetic Waves 75

Challenge: What Keeps Electrons from Falling into the Nucleus?...... 77

Describing the Forces... 79

Charge-Force is Spherical.. 79

Magnetic Force is North-South Oriented... 79

Charge-Force decreases by 1/distance-squared, Magnetic Field Strength decreases 1/distance-cubed. 80

Magnetic-Force decreases by 1/distance-cubed plus angle versus north-south, Force decreases by 1/distance-squared at north-south and 1/distance-cubed at 90 degrees. 84

Charge-Force begins the reduction at the edge of the particle....... 85

Magnetic-Force begins the reduction at each of the poles. 86

Newton Discovered Both Gravity and Integrals, and this Extension of his Gravity Formula Applies Both of Newton's Discoveries................. 91

Magnetics of Nucleus Particles in Chains Causes Enough Magnetic Attraction to Balance Charge Repulsion 94

A Magnet Stays Strong if in a Chain 95

Magnetic Neutrons are the Required Separators Extending Magnetics While Charge Decreases 100

Calculation of the Strength of Charge and Strength of Magnetic Field at Various Distances of the Radius of a Neutron 101

- Calculation of Magnetism by Comparison to Charge at Bohr Radius 103

- Measuring Charge-Force versus Magnetic-Force at Distance in Nucleus 104

'Mass' Gets Replaced by a Surface Integral Over the Electron Shell Radius 108

- Number of Nucleus Particles is Non-Changing 109

- Observed Mass is Directly Based upon the Sum of Protons and Neutrons 109

- In Every Atom, Electrons Settle into positions Both Outside and Inside the Bohr Radius 110

- The Total Number of Nucleus Particles Directly Correlates with the Atomic Radius. 111

Configurations of Nucleus Chains/Rings 113

- Basic Chain 113

- Secondary Understanding: There can be two or more neutrons in a row. 120

Magnetics of Entire Nucleus Structure Are Important and Different than Particle Magnetics 121

- Single-ring: 124

- Double-ring: 125

Building Higher Elements Needs Neutron First to Separate for Added Proton 126

- In Multiple-Ring Structures, Extra Neutrons Keep the Protons Separated 128

Structure Prefers Same Number of Particles on Every Layered Ring 129

Charge Repulsion Causes Need for Extra Neutrons in Various Isotopes – Building Larger Elements 134

Why use Magnetics versus Charge (kQQ)? 137

Why the extra Neutron in some elements? 138

Why are certain large atoms radioactive? And why consistent decay? 141

Visualization of Nucleus Chained/Ring Configurations for Various Elements 143

Multi-Layer Rings Based upon Number of Particles (by Element) 148

Isotopes and Isoforms 150

Proposal: Transparency and the Nucleus Structure – 006-C Carbon as Coal versus Diamond 152

Fundamental Question: What is a Neutron? 155

How do a proton and electron bond into a neutron at very low temperatures? 155

Glossary 156

Four Arno Vigen Science Postulates: 157

Endnotes 174

Challenge: For Newton's and Einstein's Formulas to Work Well, 'Mass' Cannot Keep Changing!

Big Hugs! Let's get going.

We have two major formulas which use mass: the gravity equation of Newton and the Energy equation of Einstein.

(1)

Newton's Gravity

$$F = G\frac{m_1 m_2}{r^2}, \qquad G = 6.67 \times 10^{-11}$$

(2)

Einstein's Energy

$$E = mc^2$$

Both use mass. Both have been tested extensively. They are established facts.

Yet, mass is something understood only intuitively. It is how much there is in Newton's apple as it is pulled towards the huge 'mass' of the Earth. We know something weights more.

Observed Mass Works – That is not the Problem

Yes, the 'observed mass' works. It is a powerful concept – 'except for the exceptions' we will resolve herein.

In the end, 'observed mass' is a simplification of two other concepts – 'observed mass' is a) the electromagnetic force of the particles as arranged, including b) that 'observed mass' force is net of a separation reduction over the electron shell/field as arranged. There is a maximum total force, which, in the Arno Vigen Scrunched Cube (AVSC) Atomic Model, the magnetic energy of the nucleons, but that energy is already used to create the electron shell, so the extra Einstein Energy and gravity is that net-force, the amount left over for distant objects.

The challenge is the replacing the 'observed mass' concept through two complex steps. First, the known Coulomb charge will connect to the nucleon magnetic force (which I will call nucleomagnetics), and second then make a clear link between those two linked forces to gravity constant, mass, and Einstein's Excess Energy using the electron shell volume reducer.

Not that Complex

In this postulate, I will have to make things more complex to prove it, and get to the simple solution at the end. My apologies! However, the concept does not require difficult math. It requires understanding just the two simple concepts. I will leave the difficult math to the back part of the postulate and footnotes.

While electrostatic particles attract opposites (positive charge (+) to negative charge (−) attracts), nucleomagnetics do the opposite. They are true Newtonian opposites. However, the fields have very different shapes; we know that funny magnetics field shape. That is the core physical reason that electrons 'settle' into specific shells and subshells. All atoms are the stable balancing point of electrostatic charge and nucleomagnetics for their nucleus of protons and neutrons and the electrons in the field.

'Observed Mass' is the concept that those electrons are in the shell, and so at any reasonable distance, the net-force has been reduced by that volume where all particles are enclosed. That is 'observed mass' and a universal continuous force called gravity.

'Mass' is a Shortcut

At NASA and JPL here in the USA, they use a shortcut for calculating exactly where to send rockets. They calculate 'mass' using the gravity force of an object at the surface of the planet. They do not calculate the gravity force of all the atoms; instead they use the surface gravity and pretend that it all sits at a point at the center.

Of course, NASA must make course corrections because that 'mass' concept misses the small complexities of the volume really is at different distances within that planet's volume.

(3)

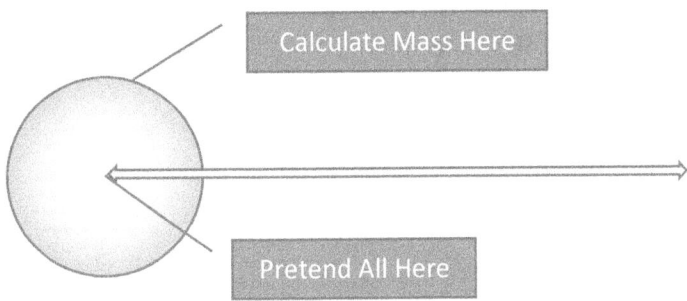

(4)

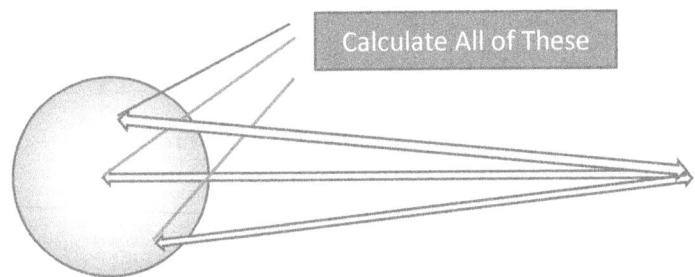

Of course, this second version is more work. It is not necessary for most problems, but for the complex situations, that extra work is needed. The 'observed mass' fix herein is that challenge one step further. I calculate 'mass' not for the atom at its edge, but for the atom for all its particles, and get the result, with the detailed separate force, if needed.

The 'fix' I propose is simply, when 'observed mass' meet challenge, use the more detailed method I describe.

My postulate is that 'mass' works great – except for the situations where it does not:

- Over a certain speed, the 'mass' changes (Lorenz).

- In molecules, 'observed mass' decreases in bonding. The 'observed mass' of two atoms is more than the 'observed mass' of their combination as a molecule.

- An electron alone supposedly has a 'tiny mass' 1,836 less than a proton.

The proof requires calculating what happens to the Electron Shell Volume, and particle positions, in each of these special situations. The postulate uses the two-part definition of 'observed mass' as a fixed 'quantitized' measure divided by a generally constant, but variable in certain situations, volume:

$$\frac{\text{Newtonian Electromagnetics Total Force}}{\text{Spacing of the Forces as Separated (the Electron Shell/Field)}}$$

And, by calculating two parts instead of one, those above situations all have the Newtonian 'AVSC Electromagnetics' Total Force remains stable, and the Volume of the Forces as Separated (the Electron Shell/Field) change to resolve the observed results.

Observed mass remains, but when required, any person can know how to do the extra work using the more complex division. Best yet, each of those situations is easily understood in 3D graphics why that volume changes the 'observed mass' shortcut. It now gives you not atoms bonding, but the specific location where the particles settle and bond, and all the forces on the particles.

(5)

AVSC Scrunched Cube Model of Atomic Shells

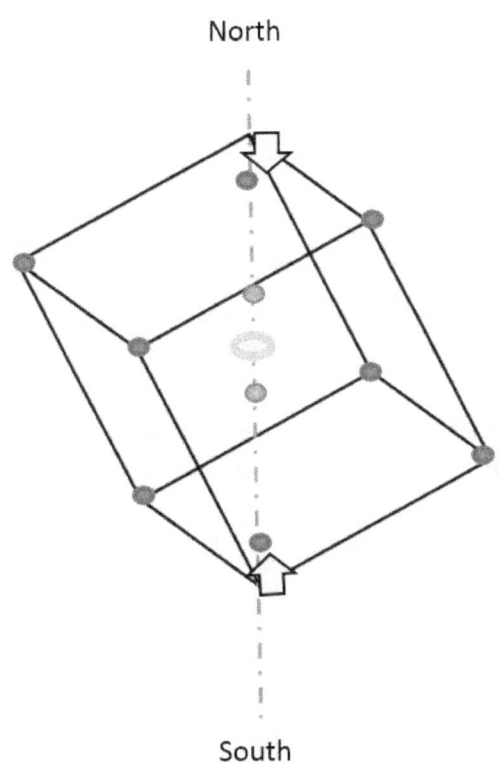

Converting Total Magnetic Force of Particles to Mass

However, there are a few challenges to 'observed mass' that become easy when you know the secret. The mass is just:

1) Magnetic Force of Atomic Particles as Arranged

2) Divided by the Volume Already in Electron-Proton Separation of the Shell-Field (with an extra -2 for the integral of the $1/x^2$ to $1/x^3$ base conversion)

(6)
$$observed\ mass = \frac{M^2(z_1 + neu_1)}{2 * \frac{4}{3}\pi(R_{ES})^3}$$

* Note that later we will discuss the potential, non-expressed portion.

You see electrons are already separated from the nucleus (specifically, the protons), so you cannot count that energy again. The electrons are already separated by that in stable atoms, so the Einstein Energy is just the rest of the energy after the basic.

AVSC Two-Part Mass is Quantized by the Number of Particles

This also will resolve quantum issues because now the base is a number of particles, so it jumps like all the quantum formulas.

Electrostatic Charge and Nucleomagnetics Forces are Not Infinite

You see, electrons and protons are electrostatic charge opposites. They really want to be together, but there is this additional energy that keeps protons and electrons apart and in shells/fields. That is the full energy less this bundle.

You cannot get them infinitely close, and get infinite energy. You hit the particle's own physical shape first.

Cause of 'Observed Mass' Not Previously Known

First, current 'observed mass' does not have a direct, understandable connection to other known particles. For over a century, scientists have struggle to find the particle that causes gravity; heck, the scientific community named a particle a graviton, but that did not seem to exist. They could not find one . . . and there should be trillions and trillions of those particles.

More recently, the scientific community have postulated a Higgs Boson particle as the source of gravity, and even went so far as to have it called a 'god' particle. Billions are spent as CERN and elsewhere, and they are lining up for the Nobel Prize because the scientists think they have a picture of one Higgs Boson particle that lasted for twenty nanoseconds.

My challenge is that the world is based upon gravity, and that cannot get based upon particles that last only a few nanoseconds.

(7)

Particle	Quantities that Exist	Life
Gravity Source	**Trillions of them**	**Consistent for centuries**
Graviton	Never found	Never found
Higgs Boson – gravity	Hundreds in a lab	Nanoseconds
Quarks	Never found	Never found

Secondly, and most importantly, mass, in current scientific literature, keeps changing. In bonding, mass changes without explanation. If moving fast, we calculate that 'mass' change by the Lorenz transformation[i]:

(8)

$$Y = \frac{1}{\sqrt{1 - \frac{v^2}{c^2}}}$$

Current thinking solves these with ingenious, insidious solutions where space and time are not fixed. However, as much as it solves some problem, it leaves others in worst shape. What is the angle at which molecules bonds? Why and how does that bond change in a magnetic field? How fast does a chemical reaction process? What are the dimensions of a quark? . . . How do I calculate anything if the 'mass' can change?

Further, that leads to science fiction. People do not age when in a spaceship. Schrodinger's Cat. Time and space warping. A warp drive. All fiction.

The postulate on the following pages provides methods for time-space to be constant and all forces to be deterministic. It gets from the number of nucleus particles, as arranged, and the electron field/shell, formally positioned, to the same Newton gravity formula and Einstein Total Energy Equation, removing the 'observed mass' from both equations entirely.

So, in current advance thinking, not only do we have base units that are not real, but also, we have base units that can change.

(9)

Force	Particles that Create	Do Base Units Change?
Electrostatic charge	Proton, Electrons	No
Nuclear binding 'strong' force	Never found	* cannot test
Electron Shell 'weak' force	Never found	* cannot test
Magnetics	Electrons moving	No
Gravity	Mass	Yes

These are the fundamental nature of the universe, and current scientific expertise has not resolved these basic questions very well. Yes, there are solutions, the Standard Model, warped space-time, and such, that resolves that by a) using stray property attributes as if discernable particles; b) making space-time flexible along with c) making mass variable. However, that is not the approach here. This solution works with non-variable time, with non-variable space, and a new definition of mass that is concrete and deterministic – based upon known particles.

Let's start with the known. Notice that two items – electrostatic charge and traditional magnetics are a) from known particles, b) have formulas that are deterministic, and from c) both the base units (protons, neutrons, and electrons) and calculations do not change.

Instead, the presentation here is the 'observed mass' is really two known items: a) the nucleomagnetics force of the nucleus particles, the protons (Z_1) and neutrons (N_1); and the average (really the facy math integral over continuous time) of the volume of the space of the particles in their electron field/shell. This occurs because the electrostatic charge of the protons decreases, so that the electrons' force expresses as stronger in every direction over time, of course. It starts 'a little bit closer' with some fancy math.

That makes the formula of 'observed mass' flow from the aggregation (integral in math) of the gravity which is really the nucleomagnetics half or the electrostatic half of electromagnetics divided by the volume used already in the basic atomic electron shell structure:

(10)

$$Gravity\ strength = \int \frac{M_{NM}^2(z_1, neu_1)}{\frac{4}{3}\pi(R_{ES})^3} = \int \frac{kZ_1\sqrt{R_{ES}}}{\frac{4}{3}\pi(R_{ES})^3}$$

Where,

m = mass

M^2 = Nucleomagnetics Force Constant

Z_1 = Number of Protons

N_1 = Number of Protons

R_{ES} = Radius of the Electron Shell

The postulate explored later is that the Radius of the Electron Shell is consistent in comparison to the protons and neutron count. That is how we get an 'observed mass' that matches to a known, unchanging count of particles. The volume increases by the added distance-cubed. The total energy from nucleomagnetics increases at that same rate; another nucleus particle creates a fixed, blog, a volume in the math language of Gauss. So, the two factors are fully linked. That transformation then allows the explanation of the changes to mass without changes to the number of particles. Something that changes the Radius of the Electron Shell will change the 'observed mass' without changing the fixed Nucleomagnetics Force from a quantitized (unit particle steps) energy of nucleus particles.

Now, it is really a much more complex in PhD math, because it is the distance of protons less the distances of the electron shell.

(11)
$$Strength = \left[C - \frac{kq_1q_1}{d^3}\right] - \left[C - \frac{kq_1q_1}{(d-R_{ES})^3}\right]$$

(12)
$$Strength = C - \frac{kq_1q_1}{d^3} - C + \frac{kq_1q_1}{(d-R_{ES})^3}$$

(13)
$$Strength = \frac{kq_1q_1}{d^3} - \frac{kq_1q_1}{(d-R_{ES})^3}$$

(14)
$$Strength = \frac{kq_1q_1[(d-R_{ES})^3 - d^3]}{d^3(d-R_{ES})^3}$$

(15)
$$Strength = \frac{kq_1q_1[d^3 - 3d^2R_{ES} + 3d(R_{ES})^2 - (R_{ES})^3 - d^3]}{d^3(d-R_{ES})^3}$$

(16)
$$Strength = \frac{kq_1q_1[-3d^2R_{ES} + 3d(R_{ES})^2 - (R_{ES})^3]}{d^3(d-R_{ES})^3}$$

(17)
$$Force = \oint \frac{kq_1q_1[-3d^2R_{ES} + 3d(R_{ES})^2 - (R_{ES})^3]}{d^3(d-R_{ES})^3}$$

And, if you can resolve that formula and integrated, you are a first-class mathematician with a job at MIT, Caltech, or JPL waiting.

For most people, you can skip that formula proof, and just understand the concept, and the final calculation, which is understandable without the fancy, complex mathematics. When you do the mathematics, the constants (C) cancel each other, and the d-cubed (which in integral becomes d-squared factors) generally cancel. Eventually, you end with a Force of 2 times the volume of that decrease space for a reduction of $2 * \frac{4}{3}\pi(R_{ES})^3$ form the basic volume with an ugly tail ('T'), the PhD full solution, if you are NASA calculating flights to Mars or beyond. It has a -1/2 integral, and . . . never mind, a few PhDs will tear my calculation apart for the next decade and figure out that I forgot a 3, or a 2, or maybe 3/2, someplace.

However, the components that lead to that direct relationship are novel. It is a long journey to get this new understanding of all the relationships 'observed mass' that resolves all the exceptions.

When needed replace mass with the two parts:

(18)

$$mass = \frac{M_A(Z_1 + N_1)}{2 * \frac{4}{3}\pi(R_{ES})^3} = \frac{kZ_1\sqrt{R_{ES}}}{2 * \frac{4}{3}\pi(R_{ES})^3}$$

I know that covers an enormous breadth, so I hope you are ready for an adventure.

Big hugs, let's keep going.

Hypothesis: All Energy comes from Electromagnetics – Electrostatic Charge and its AVSC discovered Newton Complement, Nucleomagnetics

The Arno Vigen Scrunched Cube (AVSC) Atomic Model builds all fundamental forces from a) the known particles of protons, electrons, and neutrons; and b) the forces of electrostatic charge and nucleomagnetics (the new discovery of magnetics at the particle level). It explains the connection of all fundamental forces currently defined as separate:

(19)

Arno Vigen Scrunched Cube (AVSC) Fundamental Forces:

Force	Particles that Create	Do Base Units Change?
Electrostatic charge	Proton, Electrons (but not Neutrons)	No
Nucleomagnetics	Protons, Electrons, and especially Neutrons	No
Nuclear binding 'strong' force	Protons (Z), Neutrons in chains/rings Z-N-Z-N	No
Electron Shell 'weak' force	Protons, Electrons, and Neutrons at their most basic balancing points	No
Gravity	Protons, Electrons, Neutrons (in sets at a distance in conjunction with 'weak' force R_{ES} above)	No

Which leaves two derivative forces:

(20)

Force	Particles that Create	Do Base Units Change?
Traditional Magnetics 'Motomagnetics' (nucleomagnetics in motion)	Magnetics Fields of Protons, Electrons, Neutrons moving	No
Electricity (electrostatic charge in motion)	Electrostatic Fields of Protons, Electrons, Neutrons moving	No

In my previous set of postulates, I presented that electrostatic charge and nucleomagnetics describes all the fundamental forces. The other classic fundamental forces really are combination of the two highly-integrated sets for specified distances, combinations, and configurations.

Electrostatic Charge and Nucleomagnetics Are Newtonian Opposites

Between an electron and the proton in the nucleus, electrostatic charge force is attractive (red below). Between and electron and the protons and neutrons in the nucleus, the nucleomagnetics force is repulsive. The magnetics of nucleus particles to the electrons, in electromagnetics, is the 'weak force'.

(21)

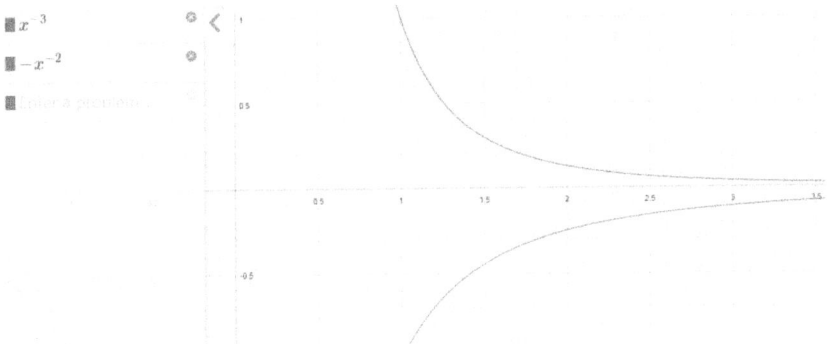

However, charge is spherical and reduces at 1/distance-squared, and nucleomagnetics has a funny shape and reduces at 1/distance-cubed. You can see the nucleomagnetics ($1/x^3$) gets closer to zero faster. You cannot see those shapes in existing calculation methods.

Those two fundamental forces are opposites. In subatomic particles, where one pushes, the other pulls. In 'weak force' holding electrons in fields/shells (between the nucleus and its electrons), electrostatic charge force pulls and nucleomagnetics force pushes. In 'strong force' (binding sets of protons and neutrons in a nucleus), it is quite the opposite; nucleomagnetics charge force pulls (holds the nucleus together in a chain/ring) and electrostatic force pushes (as nuclear decay, or in consistent large quantities as a nuclear reaction).

Yet, those two interact as quantitative-force opposites does not mean they are the same in other properties. It just means that it satisfies Newton's Third Law "For every force, there is an equal force in the opposite direction." A rubber ball and a wall interact perfectly by Newton, but when the interaction occurs at an angle, the ball goes off in a different direction, it squeezes and deforms, and the wall does not deform. The differences include a) strength-over-distance, 1/distance-squared (electrostatic) versus 1/distance-cubed (nucleomagnetics); and b) orientation, spherically consistent (electrostatic) versus two hemispheres with strength by the inclination angle from the poles (nucleomagnetics).

That net of the different types creates a net force over distance where the two balance each other at the Bohr radius as modified by the nucleomagnetics inclination/longitude angle. At the poles, the calculation for 01-H Hydrogen is the Bohr radius because that is where electrons settle – at the weakest repulsion force location. It also explains 02-He Helium as full (2 poles). In larger atoms, the forces can push differently – depending on that nucleomagnetics inclination angle of the full set of electrons which settle off the nucleomagnetics polar axis. That below is adding the electrostatic and nucleomagnetics forces of a subatomic particle for a particular distance (in the direction along the nucleomagnetics pole for illustration).

(22)

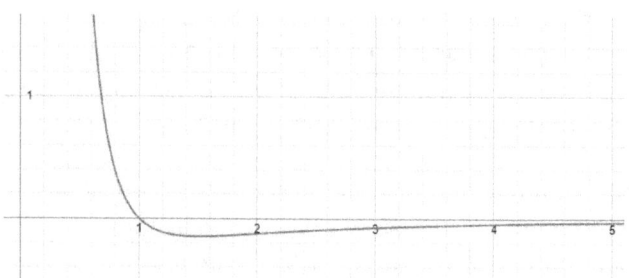

1) The nucleus is held together by the 'strong force'; that is really the magnetic chain of nucleus particles, protons then neutrons, then protons, and so on; that chain is magnetics, and a magnetic chain does not decrease in strength along its length, so a nucleus magnetic chain, ring, or more complex structure is stronger than electrostatic repulsion of the protons in the chain from each other. The key is proton separation. The proton-proton repulsion reduces at 1/distance-squared enough that the magnetic nucleus chain/ring is very, very stable. Remember that an extra neutron in the structure adds to the stability.

(23)

Side View

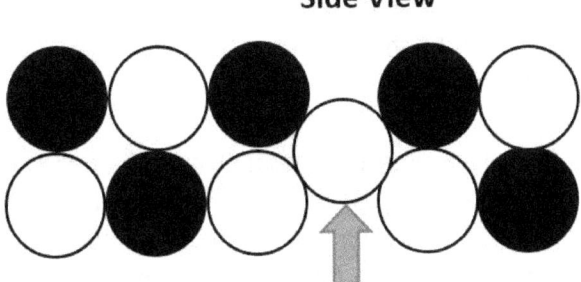

When the neutron is not there, or disturbed, the protons get too close, and the nucleus will decay, and if that creates a chain of decay where the output of one decay trigger another atomic decay, that is nuclear energy or a nuclear bomb.

(24)

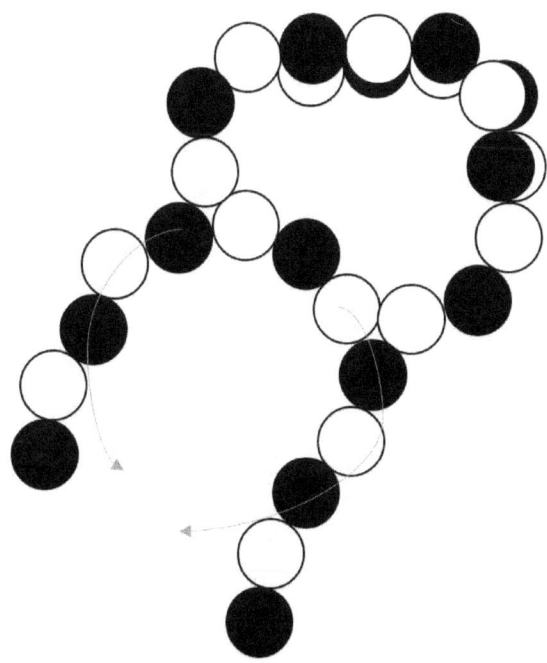

2) The second fundamental force is 'weak force'; it is reality that an electron does not fall into the nucleus, even though all electrons should collapse because their negative (-) electrostatic charge is attracted to positive (+) charges protons in the nucleus.

That electron settling position, away from the nucleus, is the balancing point of the electrostatic versus nucleomagnetics forces. For electrons in the 'shell' or field around protons in the nucleus, this is the point where the electrostatic force attraction equals (balances) the repulsion of nucleomagnetics (when that is a proton-to-electron interaction). The 'z=1, x=0, y=0' (or its spherical equal) intercept is the balancing point, the electron shell radius, and generally the Bohr's radius.

When you add the two forces, electrostatic charge and nucleomagnetics:

(21)

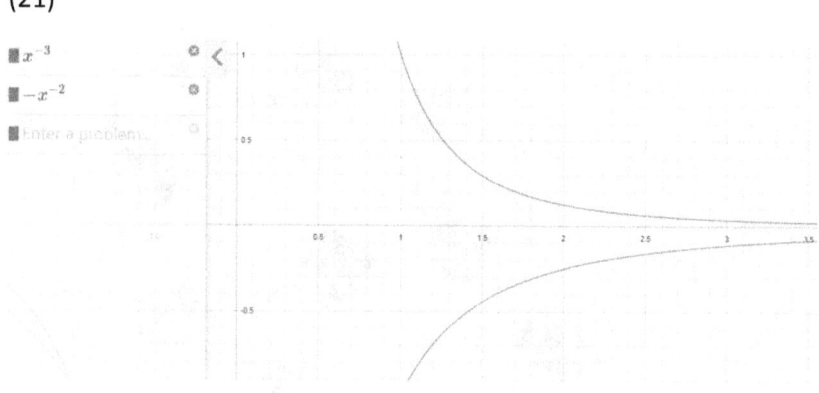

You get a net-force of nucleomagnetics, and the 'weak force' holding electrons in a shell/field.

(22)

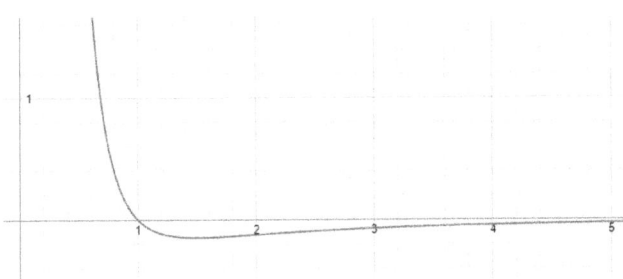

In fact, this is an amazing graph. In physics, a positive is repulsive, and a negative is attractive. That means that at the downward crossover point, when the particle gets closer, it gets repulsed (back to the balance point). When the particles move further apart, the particles are attracted (negative force versus their distance separation). Any physic-force graph with this positive to negative crossing of zero is very, very special.

(25)

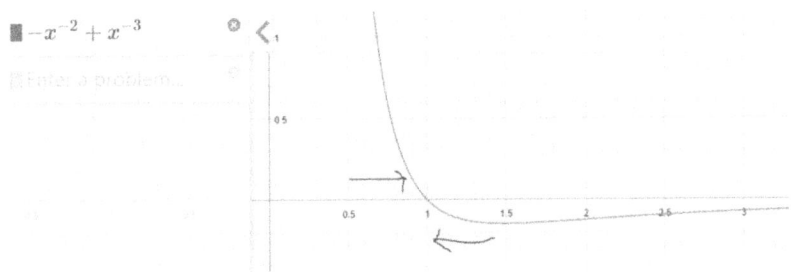

In pictures, that is any physics force graph where a force function starts positive, crossing zero to negative. When electron-proton force balance with a tendency for that balance keeping the particle at that distance – 'in a shell'.

3) Molecular bonds occur when the same electron can hold a position in two different atoms. That is, when both atoms have an open settling position, and other criteria ensue (such

as, other electrons from those atoms do not push the rest of the structure apart); this is the basic physics and chemistry that generates all molecular bonds. Every atom and molecule has a structure (relative to its set of all its subatomic particles). The nucleomagnetics make the electron shells and subshells hold at this balancing distance, as easy to understand 3D 'scrunched cube' model.

It is the electrons held outward, but at different distances that makes the Periodic Chart of Elements. The first two electrons (traditionally called Subshell-s, in AVSC called Subshell-m for the magnetic axis) get 'scrunched' so that the cube become flattened, like an accordion, slightly.

(26)

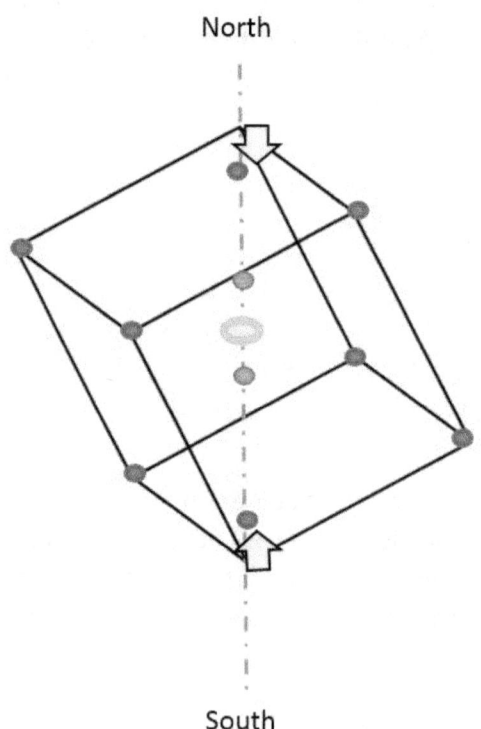

By way, the 'scrunch' is the lower nucleomagnetics repulsion creating two electrons at a closer, lower repulsion position – consistent with the current Periodic Table of Elements.

Atoms build from the center in two hemispheres. The electrons like to be in opposite positions in any structure; as both are electrostatic negatives, electrons want to settle as far away from each other as possible. The first two go to the poles (blue 2m1, 2m2 electrons formerly named 2s2). Then, layers ('shells') build 2x1-squared, 2x2-squared, 2x3squared, interweaving to create the $2x1^2$, $2x2^2$, interweaved $2x2^2$, $2x3^2$, interweaved $2x3^2$, $2x4^2$, interweaved $2x4^2$ shells.

The picture shows the electron 'settling' positions for a full Shell-2 atom.

One can see the interweaving when Shell-3 is added to the picture. An extra layer finds positions between Shell-2 without growing to the next $2xN^2$ size.

(27)

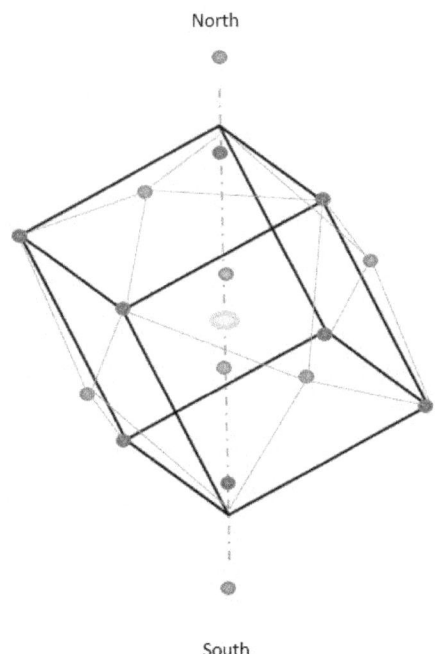

North

South

Bonds are not *(well, except in certain true double, shared bonds of molecules like O2)* a shared electron pair as described by Pauli, but single electrons filling nucleomagnetics positions in two (2) different directions relative to the electrons towards those two (2) atoms. It is a deterministic 3D structure for every atom, for every molecule, and a 3D animation for every chemical reaction *at the particle level*.

(28)

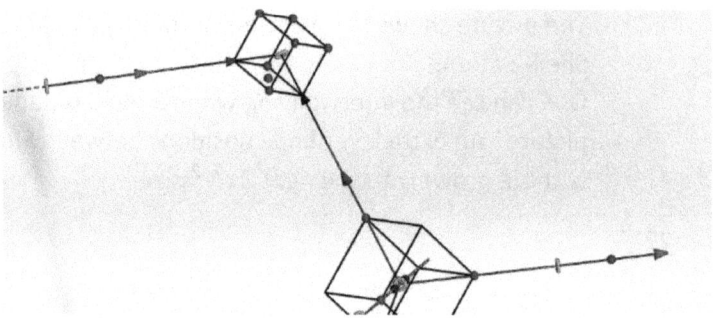

We can draw and calculate, not just the bond angles, but the position, forces, and orientation for every particle, and how they will react in 3D animation. This leads to not just the relative energy levels, but also the chemical reaction speed.

Pauli exclusion gets replaced by the fact that in a spherical energy balance, the next (even) electron almost always wants to be exactly opposite the unbalanced (odd) electron positions in a sphere.

Further, there are AVSC exceptions to the aufbau electron filling rules. All atoms have a settling position for each electron; the nucleus and electrons move at a set – creating non-independent 'orbits'. This creates molecular sets with actual relative positions.

Of course, Pauli normally works well. So, the main exceptions to Pauli in AVSC are the AVSC equatorial electron configurations. In 29-Cu Copper, equatorial electrons create high electrical conductivity (very low resistance) in a manner that can get calculated by AVSC.

In the AVSC 3D model, you can see that three electrons, 4e1, 4e2, and 4e3, settle away from the other structure, and so they can enter and exit with little interference. This compares to the other electrons that would need to get around the fields of other subshell electrons in order to flow as freely as electrons flow in the AVSC Subshell-3e model. Better yet, the calculations are now supported by a 3D engineering model.

(29)

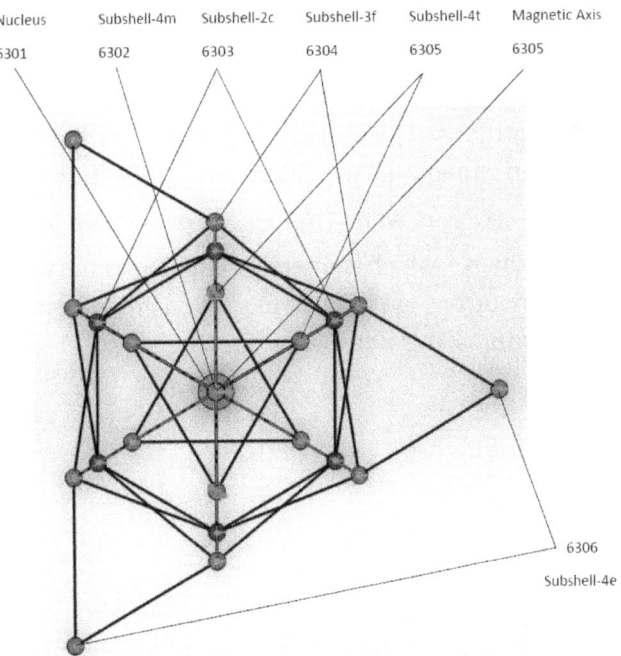

Further, one of the corollaries is that may Hydrogen atoms in molecules sit Proton-Outward – creating adhesions and cohesion.

(30) It makes the Hydroxide attached to an atom look like the upper left section of the CAD/CAM drawing below:

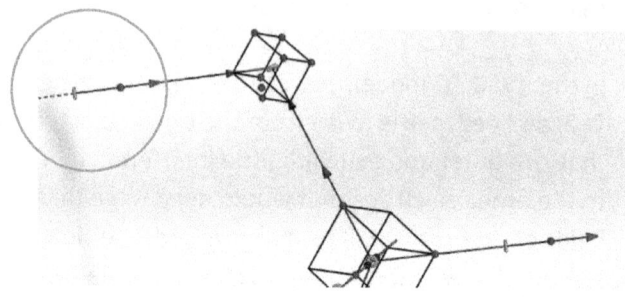

Who has ever heard of that? An atom where the proton settles outward, outside the electron. It is critical for all Chemistry in adhesion and cohesion calculations.

4) When the spread of electrons gets closer than the nucleus protons. The proton electrostatic (+) strength has decreases in strength by the volume of the electron shell (which is $\frac{4}{3}\pi(R_{ES})^3$ when (R_{ES}) is the average radius of the electron shells – which is directly matching to the count of protons and neutrons in the nucleus. When integrated over millions of atoms, this net-force is gravity.

Relating Electrostatic and Nucleomagnetics Force

Remember that the Radius of the Electron Shell (R_{ES}) is the balancing point of two equations. In stable atoms, those both balance at that special distance, (R_{ES}).

(31)
$$Electrostatic\ Charge\ Force = \frac{'kQQ'}{R_{ES}^2}$$

(32)
$$Nucleomagnetics\ Force = \frac{M(Z_1 + N_1)}{\frac{8}{3}\pi(R_{ES})^3}$$

For the electrons in the shell/field then, we can be a balancing point:

Now, the 'k' from the original Coulomb's Law is really a simplification the AVSC Electron Shell (ES) volume version, the formula have the Coulomb's k versus a k_{ES} where I reduce the 4/3π (much like Planck and reduced Planck):

(33)
$$\frac{k_{ES}QQ}{\frac{4}{3}\pi R_{ES}^2} = \frac{M_{ES}(Z_1 + N_1)}{\frac{4}{3}\pi(R_{ES})^3}$$

(34)
$$\frac{kQ_1Q_2}{R_{ES}^2} - \frac{M(Z_1 + N_1)M(Z_2 + N_2)}{R_{ES}^3} = 0$$

(35)
$$\frac{kQ_1Q_2}{R_{ES}^2} = \frac{M(Z_1+N_1)M(Z_2+N_2)}{R_{ES}^3}$$

So, using the original Coulomb constant (without the π), I separate the forces into the two sides one atom's particles and charge (Z_1, N_1, Q_1) times the other atom's particles and charge (Z_2, N_2, Q_2), so:

(36)
$$\frac{M(Z_1+N_1)M(Z_2+N_2)}{R_{ES}^3} = \frac{kQ(Z_1)Q(Z_2)}{R_{ES}^2}$$

(37)
$$\frac{M(Z_1+N_1)M(Z_2+N_2)}{R_{ES}^3} = \frac{kR_{ES}Q(Z_1)Q(Z_2)}{R_{ES}^3}$$

(38)
$$M(Z_1+N_1)M(Z_2+N_2) = kR_{ES}Q(Z_1)Q(Z_2)$$

I will make a major assumption that for tow particles equality, the two multipliers themselves are also equal,

So,

(39)
$$M(Z_1+N_1)M(Z_2+N_2) = Q(Z_1)\sqrt{kR_{ES}} * Q(Z_2)\sqrt{kR_{ES}}$$

Becomes

(40)
$$M(Z_1+N_1) = Q(Z_1)\sqrt{kR_{ES}}$$

Think about it. In an isotope, the 'kQQ' does not change as the number of protons is the same $Q(Z_1)$, but the nucleomagnetics particles of $M(Z_1 + N_1)$ changes by the additional neutron(s) (N_1).

R_{ES} Changes in Direct Relationship to Nucleus Particle Count

That means that for measurements of stable atom's force on each other, at a distance, the Magnetics Force, the 'observed mass' will operate best when using the Magnetic Force factor (Atomic Weight), and not the Atomic Number.

(41)
$$\frac{M(Z_1 + N_1)M(Z_2 + N_2)}{(R_{ES})^3} = \frac{kQ_1Q_2}{(R_{ES})^2}$$

(42)
$$\frac{M(Z_1 + N_1)M(Z_2 + N_2)}{(R_{ES})^3} = \frac{kR_{ES}Q_1Q_2}{(R_{ES})^3}$$

The left side makes Magnetics match to the volume. Their change is direct. However, the right side has does not change directly.

Because the direct relationship goes with nucleomagnetics, observed mass prefers that measure

Yet, at a distance, the calculation is 1-distance-square v

I reality, there is both, but nucleomagnetics is immaterial after one meter. At that distance, nucleomagnetics is 1/100,000,000,000x smaller. At the moon, it is immaterial.

Because the overpowering force at a distance is electrostatic charge, the gravity calculation uses 1/distance squared.

If we use a measure of the Protons-only (the Atomic Weight), then we would get the Radius of the Electrons Shell changes not consistently.

For Carbon,

(43)
$$\frac{6}{1} = \frac{(6+6)}{R_{ES}} = \frac{12}{R_{ES}}$$
$$\frac{R_{ES}}{1} = \frac{12}{6} = 2$$

Adding just one Proton, for a low-isotope of Nitrogen,

(44)
$$\frac{7}{1} = \frac{(7+6)}{R_{ES}} = \frac{13}{R_{ES}}$$
$$\frac{R_{ES}}{1} = \frac{13}{7} = 1.85$$

So, the radius will not be consistent based upon Proton-only count – **even though** the reflected force at a distance is the 1/distance-squared electrostatic charge differential.

Notice that this is Average Radius of the Electrons Shell – not Covalent Bonding Distance, not Edge of Molecule, not Outer Shell

All the existing tables of electron shells focus on the most outer shell,

- that is the position for ionization

- that is the position for bonding

Complexity #1: If you try to review the covalent radius (used for bonding) or Van der Waals radius (used for gas, liquid solid calculations, you will not find these radios apply.

Instead this is the mathematical average radius of the various.

1) As molecules gain in nucleus particles (like 86-Pb Lead), the location of the first subshell (1m2) gets closer.

2) That allows the outer shells to get further out, so long as the average become as per the calculation (Z+N)x.

3) The radius become bigger, but not much.

(45)

$$R_{ES} = \frac{(Q_1)^2 k}{M(Z_1 + N_1)^2}$$

That radio of Atomic Number (kQQ)/ Atomic Weight (Z+N) is 1:2 after Hydrogen, then change until about 1:2.6 in heavy radioactive atoms. That means the radius of atoms remains in the same range, every if as the particles go from about 4 to 250, the radius goes from 2 to only 2.6.

In fact, this is one of the critical reason for the ideal gas law, a 'mole', and for Avogadro's number. More and more packs into the same space. A gas of a certain space and temperature can have X-Times the volume, and the same number of atoms, even with huge differences in the 'observed mass' of the individual atoms or molecules.

Complexity #2: Remember that some of those electrons are not at the nucleomagnetics axis which means they have the $(1+3SIN(\theta)^2)$ factor applied to the nucleomagnetics off-axis repulsion portion.

Complexity #3: Remember temperature (rotational forces on the atoms and molecules) and external magnetic fields also change these calculations.

The complexities of a surface integral versus the integral of the specific multiple radius positions will not get addresses herein. It is the level of discussion for the top Mathematics PhDs.

Root-G and Root-k Special Uses

And then that applies from two ends, so the Newton Theorem is:

(46)
$$F = \frac{\sqrt{G}(m_1) * \sqrt{G}(m_1)}{d^2} = \frac{G^1 m_1 m_1}{d^2}$$

I know that splitting the gravity constant to apply it as the square root to each of the factors is not often done. However, that is one of the tricks that will make the path to the revised calculation clear, and convert to a formula where know particles and their direct electromagnetic fields comprise the formula.

That makes the charge force the netting of proton force and electron force with the electrons just that much closer (on average):

(47)

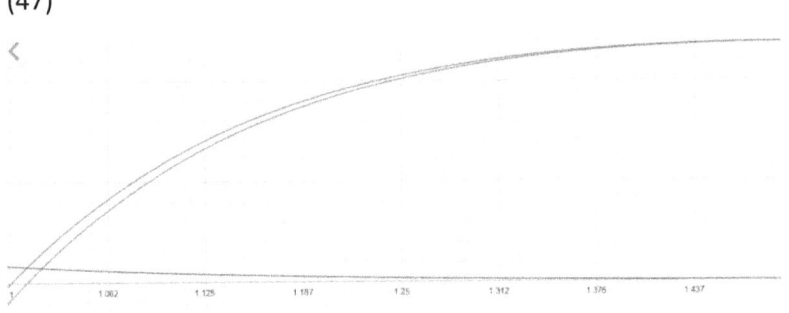

Atom-Gravity (48)

Factor	Base	Field Strength decreases by	Radius
Nucleus creates magnetic field force	Protons + Neutrons = 'Nucleus Mass'	While complex to calculate specific electron balance points within a magnetic field[ii], the total field balance is simple: (Magnetic Repulsion = Charge Attraction) over / Volume of (Shell-Radius)³	Nucleus
Electrons has closer Charge force	Net Charge	$$\frac{kQ_E}{(d \mp r)^3} - \frac{kQ_P}{d^3} = 0$$ Excess Charge of Electrons less the Charge of Protons averaged over time over the electron shell positions	Electron Shell**
Distant body recognizes only the 2nd larger element	Product with the Electron Shell's Charge as the 'math eliminated' connection	The Charge Attraction cancels to: $$\frac{M_A(Z,N)_1 M_A(Z,N)_2}{d^2(\frac{8}{3}\pi R_{ES}^3)}$$ Which is the same as Newton $$\frac{Gm_1 m_2}{d^2}$$	Galaxies away

* I use Q versus $q_1 q_2$ because this is the fundamental of one Charge attraction or repulsion to one other Charge. It is not a variable, but 1:1.

** The electron shell distance is the combination of the electron-electron repulsion (spherical) plus the nucleus magnetic field (north-south oriented). However, for electron repulsion, there is the exactly balancing number of proton attraction, so in the end for the average the driving element is just the nucleus magnetic field.[iii]

Gravity by Arno

Nucleus particles magnetics (nucleomagnetics) cause electron-shell-distance

Nucleus Particle Magnetics Caused

Gravity is a relayed force.

Relayed

Gravity is a net force. The electrostatic charge-at-a-distance force of electrons is greater than a) Electrostatic Charge-at-a-distance force of the nucleus protons, even with b) lower Electrostatic Charge of electrons on the far side of any electron-orbits -- both because of 1/distance-squared.

Net-Charge-$1/d^2$ Force

Gravity is an integral over time. It is not a separate particle.

Derived integral over time

Gravity is a nucleus-particle-magnetics-caused, electron-shell-relayed, net-electrostatic charge-$1/d^2$-force calculated by an integral over time.

Calculation of Gravity from Charge versus Newton $6.674 \times 10^{-11} \frac{m^3}{kg(s^2)}$

These Electron-Shell Separation has many factors which will get handled extensively in the footnotes. There is a lot of moving parts, many of which need further validation for a 'published paper' level of acceptance. However, in general, you can see these implications of these factors in the below example of 001-H Hydrogen to Hydrogen at 1-meter apart:

- The Charge Force (if no offset by electron) is significant - 10^{-28}
- The Gravity Force (using Newton) is 'tiny' - 10^{-62}
- The Net-Electrostatic Charge Force (Arno Vigen method) is a similar 'tiny' - 10^{-62}

The Gross-Charge Calculation for 001-H Hydrogen electron (-) to 001-H Hydrogen (+) atom at a distance of 1-meter is as follows:

(49)

Factor	Net-Charge Calculation
Charge Force factor	$k = 10^{10}$ m² / (s²) [9.03]×10⁺⁹]
Charge of orbiting Electron	$Q=10^{-19}$ [1.602]×10⁻¹⁹]
Charge of distance Proton	$Q=10^{-19}$ [1.602]×10⁻¹⁹]
Distance	d=1 m or 10^0
Exponent shortcut	+k+Q+Q-d-d
Gross Charge Short-cut calculation	10-19-19-0-0 = -28
Gross Charge Force	10^{-28} m¹ / (s²)

The Net-Charge Calculation for 001-H Hydrogen atom to 001-H Hydrogen atom at a distance of 1-meter is as follows:

(50)
$$\frac{M_A(Z,N)_1 M_A(Z,N)_2}{d^2(\frac{8}{3}\pi R_{ES}^3)} = \frac{kQQ}{d^2(\frac{8}{3}\pi R_{ES}^3)}$$

(51)

Factor	Net-Charge Calculation
Charge Force factor	$k = 10^{10}$ m² / (s²) [9.03)×10^{+9}]
Charge of orbiting Electron	$Q=10^{-19}$ [1.602)×10^{-19}]
Charge of distance Proton	$Q=10^{-19}$ [1.602)×10^{-19}]
Distance	d=1 m
Radius of the Nucleus (protons and neutrons)	r=10^{-11} m (Bohr radius which is 5.27 x 10^{-11})
Exponent shortcut	+k+Q+Q-d-d
Gross Charge Short-cut calculation	10-19-19-0-0 = -28
Net Electron-Shell Charge Short-cut calculation	-28-11-11-11-1 = -61
Net-Charge Force of the Electron (protons and neutrons) New AVSC Formula	10^{-61} m¹ / (s²) /($\frac{8}{3}\pi$) = 10^{-62} m¹ / (s²)

This is just the basic force of charge force as pushed out by the electron-shell. The gross charge of the electrons 10^{-28} gets reduced by the $\frac{8}{3}\pi R_{es}^3$, which is another 10^{-34} creating a net-charge for distant objects of 10^{-62}.

How does that compare with the standard calculation of gravity using Newton's formula?

Gravity from 001-H Hydrogen atom to 001-H Hydrogen atom at a distance of 1-meter using the Newton method is:

(52)

Factor	Gravity Estimation
Gravity Constant (Newton)	$G = 10^{-10}$ m^3 kg^2/ (s^2) [6.67)×10^{-11}]
Mass of close 001-H Hydrogen	$m=10^{-26}$ kg [1.602)×10^{-24}]
Mass distant 001-H Hydrogen	$m=10^{-26}$ kg [1.602)×10^{-24}]
Distance	d=1 m or 10^0
Exponent shortcut	+G+m+m-d-d
Short-cut calculation	-10-26-26+0+0=-62+0=-62
Newton Gravity Force	10^{-62} m^1 / (s^2)

The new AVSC calculation gets to the same range of strength; net-charge at electrons shells matches Newton's gravity force for a simple 001-H at 10^{-62} m^1 / (s^2).

(53)

Net-Charge Force (AVSC)	10^{-62} m^1 / (s^2)

Total Energy with Nucleomagnetics

This new model changes the calculation of total energy because of the extra energy in the shape of the nucleomagnetics force.

In the basic calculation of energy, we have the minimum energy for balancing nucleomagnetics and electrostatic force (plus a portion of that field that is a potential-only because of orientation):

(40)

$Force$

$$= \frac{kq_z q_e}{d^2} + \frac{M(z,n)\left(1 + 3SIN(\theta_{z,neu})^2\right)^{\frac{1}{2}} * M(e)(1 + 3SIN(\theta_e)^2)^{\frac{1}{2}}}{d^3}$$
$$+ T'$$
$$+ i \left(\frac{M(z,n)\left(2 - \left(1 + 3SIN(\theta_{z,neu})^2\right)^{\frac{1}{2}}\right) * M(e)\left(2 - (1 + 3SIN(\theta_e)^2)^{\frac{1}{2}}\right)}{d^3} \right)$$

For the moment, we will ignore the non-expressed potential portion of the equation. In that way, we can find the Nucleomagnetics Constant from the simplest case, the 01-H Hydrogen atom 0-neutron isotope, and ignoring the tensor for the moment:

(40)

At $d = R_{ES}$,

$$0 = \frac{kqq}{(5.49E-11)^2} + \frac{M^2(1)(1 + 3SIN(0)^2)^{\frac{1}{2}} * (1)(1 + 3SIN(0)^2)^{\frac{1}{2}}}{(5.49E-11)^3}$$

Which simplifies further to:

(41)
$$0 = \frac{kqq}{(5.49E-11)^2} + \frac{M^2(1)(1+3*0)^{\frac{1}{2}} * (1)(1+3*0)^{\frac{1}{2}}}{(5.49E-11)^3}$$

(41)
$$0 = \frac{k(1)(-1)}{(5.49E-11)^2} + \frac{M^2(1)(1)^{\frac{1}{2}} * (1)(1)^{\frac{1}{2}}}{(5.49E-11)^3}$$

(42)
$$0 = -\frac{kqq}{(5.49E-11)^2} + \frac{M^2}{(5.49E-11)^3}$$

Note that for O1-H e-1m1, all the units are 1-multipliers; this is not the case for other atoms.

Now, this is resolves to

(44)
$$\frac{(2.31E-28)}{(5.49E-11)^2} = \frac{M^2}{(5.49E-11)^3}$$

(45)
$$(5.49E-11)(2.31E-28) = M^2$$

(45)
$$M^2 = 1.22E - 38$$

AVSC Base Equation and Bohr Radius Only Expressed Forces

But the above Bohr radius equilibrium is expressed forces only. This calculation assumes that the electrons is a) on the axis, and b) oriented itself along that nucleus-electron axis – which for 01-H Hydrogen happen to be the same. As a result, there is a non-real portion that remains a potential. (And, all the 1+3SIN factors are 1-multipliers.

(44)

$$0 = \frac{(2.31E-28)(1)(-1)}{(5.49E-11)^2} + \frac{M^2}{(5.49E-11)^3} + i\frac{M^2}{(5.49E-11)^3}$$

With the potential portion, we now have a triplet to describe a particles inherent force. (Think elementary particles.)

This extra energy is expressed, and known, in external magnetic fields. It changes the position and energy of particles.

As the particle changes its magnetic orientation, some of the nucleomagnetics energy moves from a potential to expressed.

In a stable 0-neutron isotope of 01-H Hydrogen atom, a) the electron is on the axis of the nucleus, and b) the electron generally orients itself (by rotation to get the lowest energy position) such that the nucleus is on the axis of the electron. Both side . . .

(40)

Force

$$= \frac{kq_z q_e}{d^2} + \frac{M(z,n)(1+3SIN(0)^2)^{\frac{1}{2}} * M(e)(1+3SIN(0)^2)^{\frac{1}{2}}}{d^3} + T'$$

$$+ i \left(\frac{M(z,n)\left(2-(1+3SIN(0)^2)^{\frac{1}{2}}\right) * M(e)\left(2-(1+3SIN(0)^2)^{\frac{1}{2}}\right)}{d^3} \right)$$

$$= \frac{kq_z q_e}{d^2} + \frac{M(z,n)(1+0)^{\frac{1}{2}} * M(e)(1+0)^{\frac{1}{2}}}{d^3} + T'$$

$$+ i \left(\frac{M(z,n)\left(2-(1+0)^{\frac{1}{2}}\right) * M(e)\left(2-(1+0)^{\frac{1}{2}}\right)}{d^3} \right)$$

$$= \frac{kq_z q_e}{d^2} + \frac{M(z,n)(1) * M(e)(1)}{d^3} + T'$$

$$+ i \left(\frac{M(z,n)(2-1) * M(e)(2-1)}{d^3} \right)$$

$$= \frac{kq_z q_e}{d^2} + \frac{M(z,n)(1) * M(e)(1)}{d^3} + T'$$

$$+ i \left(\frac{M(z,n)(1) * M(e)(1)}{d^3} \right)$$

This is very balanced, the express energy (1) has the same Energy as the potential (1).

(45)

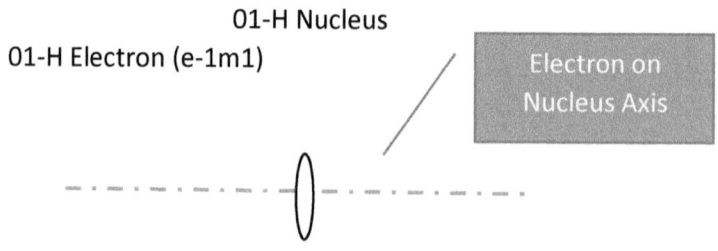

Visually, this situation is very special. All the particles must be on the axis of the other particle. That means that the MAX(potential) = 1x and the MIN(expressed) = 1x totally 2x at that special location, but the MAX(expressed) = 2x which leaves MIN(potential) = 0x which still totals 2x.

By the way, gold and silver has their outermost electrons as this 2nd special settling location.

(46)

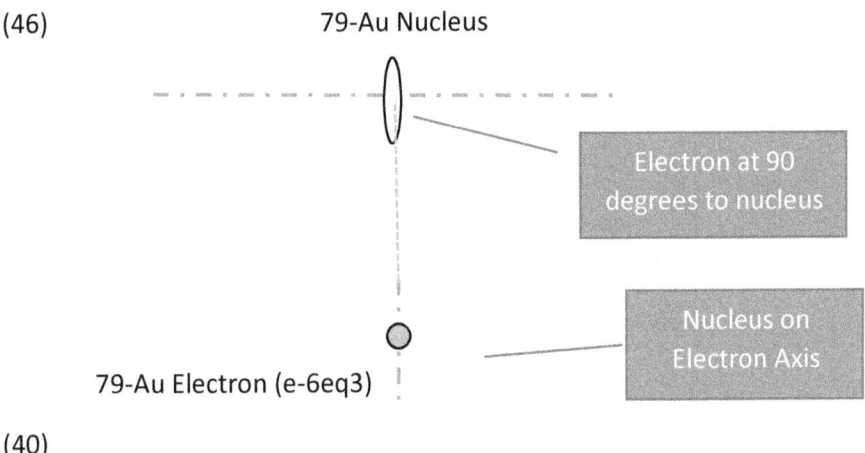

(40)

Force

$$= \frac{kq_zq_e}{d^2} + \frac{M(z,n)(1+3SIN(90)^2)^{\frac{1}{2}} * M(e)(1+3SIN(0)^2)^{\frac{1}{2}}}{d^3} + T'$$

$$+ i\left(\frac{M(z,n)\left(2-(1+3SIN(90)^2)^{\frac{1}{2}}\right) * M(e)\left(2-(1+3SIN(0)^2)^{\frac{1}{2}}\right)}{d^3}\right)$$

$$= \frac{kq_zq_e}{d^2} + \frac{M(z,n)(1+3(1))^{\frac{1}{2}} * M(e)(1+0)^{\frac{1}{2}}}{d^3} + T'$$

$$+ i\left(\frac{M(z,n)\left(2-(1+3)^{\frac{1}{2}}\right) * M(e)\left(2-(1+0)^{\frac{1}{2}}\right)}{d^3}\right)$$

$$= \frac{kq_zq_e}{d^2} + \frac{M(z,n)(2) * M(e)(1)}{d^3} + T'$$

$$+ i\left(\frac{M(z,n)(2-2) * M(e)(2-1)}{d^3}\right)$$

$$= \frac{kq_zq_e}{d^2} + \frac{M(z,n)(2) * M(e)(1)}{d^3} + T'$$

$$+ i\left(\frac{M(z,n)(0) * M(e)(1)}{d^3}\right)$$

(46)

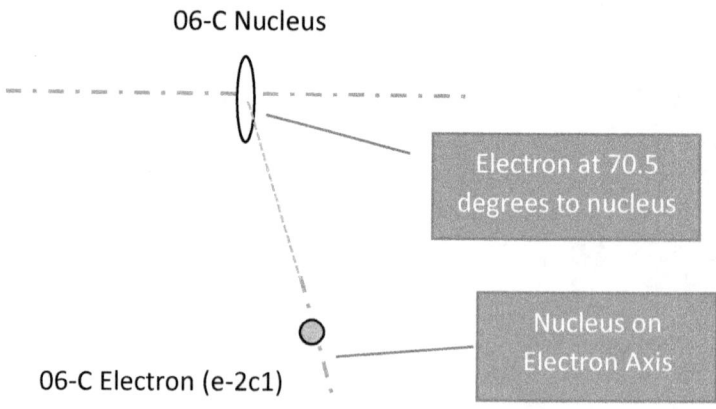

06-C Nucleus

Electron at 70.5 degrees to nucleus

Nucleus on Electron Axis

06-C Electron (e-2c1)

However, for other atoms some electrons are not on the axis, so the expressed nucleomagnetics force increases according. For a stable 06-C Carbon atom, the electron sit a 109.5 (70.5) degrees relative to the nucleomagnetics axis of the nucleus.

As a result, the expressed portion changes – more is expressed and less is a potential:

(40)

$$Force = \frac{kq_zq_e}{d^2} + \frac{M(z,n)(1+3SIN(109.5)^2)^{\frac{1}{2}} * M(e)(1+3SIN(0)^2)^{\frac{1}{2}}}{d^3} + T' +$$

$$i\left(\frac{M(z,n)\left(2-(1+3SIN(109.5)^2)^{\frac{1}{2}}\right)*M(e)\left(2-(1+3SIN(0)^2)^{\frac{1}{2}}\right)}{d^3}\right) = \frac{kq_zq_e}{d^2} +$$

$$\frac{M(z,n)(1+3*0.942641)^{\frac{1}{2}} * M(e)(1+0)^{\frac{1}{2}}}{d^3} + T' +$$

$$i\left(\frac{M(z,n)\left(2-(1+0.942641)^{\frac{1}{2}}\right)*M(e)\left(2-(1+0)^{\frac{1}{2}}\right)}{d^3}\right) = \frac{kq_zq_e}{d^2} +$$

$$\frac{M(z,n)(3.827924)^{\frac{1}{2}} * M(e)(1+0)^{\frac{1}{2}}}{d^3} + T' +$$

$$i\left(\frac{M(z,n)\left(2-(3.827924)^{\frac{1}{2}}\right)*M(e)\left(2-(1+0)^{\frac{1}{2}}\right)}{d^3}\right) = \frac{kq_zq_e}{d^2} +$$

$$\frac{M(z,n)(1.96508)*M(e)(1+0)^{\frac{1}{2}}}{d^3} + T' +$$

$$i\left(\frac{M(z,n)(2-(1.96508))*M(e)\left(2-(1+0)^{\frac{1}{2}}\right)}{d^3}\right) = \frac{kq_zq_e}{d^2} +$$

$$\frac{M(z,n)(1.96508)*M(e)(1)}{d^3} + T' + i\left(\frac{M(z,n)(2-1)*M(e)(2-1)}{d^3}\right) = \frac{kq_zq_e}{d^2} +$$

$$\frac{M(z,n)(0.43492)*M(e)(1)}{d^3} + T' + i\left(\frac{M(z,n)(1)*M(e)(1)}{d^3}\right)$$

If the Universe is in Balance, How Can There Be Overall Net Attraction?

"For every action, there is an equal and opposite reaction."

Sir Isaac Newton

You have all these net electron-shell attractions going on forever. If so, how can it be that the universe is all these attraction forces? What is the offset? Where are the balancing repulsions?

The forces do offset, but in a creative way. The force does net over all distances from the atom. However, there is a huge negative net force, a tendency for repulsion, at the shells, and thereby a leftover attractive force outside. But that leftover goes on forever (although it gets really small), so the gravity-force actually balances the electron-shell repulsion zone. However, they are opposite types – big, fat at the shells versus long and skinny everywhere else.

The shell of electrons creates this huge barrier. We see this in all gases; we see this in the limited locations of bonding and bonding angles and strengths of bonds. An atom of a particular element is highly protected by its electrons shell. When most other atoms approach, the other atom's electrons near the sitting atom's electrons first, and the atoms bounce off without any bonding.

As a result, there is a huge ring of repulsion because if other atoms, with electrons also on the outside, then as they get close, the electron repulsion overwhelms the general attraction. Another atom bounces away. It is rare and special to approach correctly to build an atom-to-atom bond that lasts.

Here is an example of the net attraction/repulsion for a 006-C Carbon atom based upon distance – in the line of one of the surrounding electrons. It has a tiny net attraction after the distance of about 2x Shell-2. It has a huge repulsion when it gets anywhere near Shell-1.

(54)

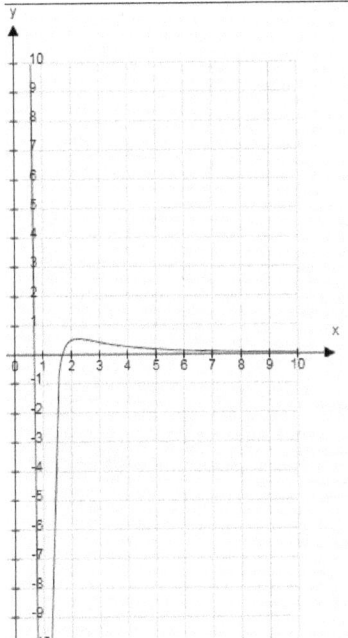

$$y = 6 \cdot \left(\frac{1}{x}\right) \cdot \left(\frac{1}{x}\right) - \left(\frac{1}{x-1}\right) \cdot \left(\frac{1}{x-1}\right)$$

Near Shell-1, the repulsion (negative) is enormous.

Attraction peaks at just over twice the distance of the electron shell.

Then attraction reduces – by 1/distance-squared, of course. There it is again, gravity is that tiny net left over, even if electrons are in the way. Gravity is something different near atoms; it is called bonding force and it is more often negative up close, and attractive far away.

At long distances, that large nucleus attraction in a direct line overcomes the electron repulsions which are distributed over the shells.

At short distances, the electron repulsions overcome the nucleus pull and that protects the stability of atom structure.

Over all distances, the world order balances. Strong repulsion nearby balances tiny attractions from distances going on forever. The two balance.

Fundamental Question: Why does 'mass' change at speeds near the speed of life?

"Mass is found to increase with velocity, but appreciable increases require velocities near that of light."[iv]

Another way that scientists have observed that 'mass' changes is for atoms moving at a material percentage of the speed of light. However, when you substitute in the deeper calculation of observed 'mass' from the above, one factor, R_{ES}, makes that 'mass' change understandable, and the other, $(Z + N)$, does not.

When moving at the speed of light, the electrons orbiting have a challenge on the back side, they have speed limit so the electrons are slow to catch the moving nucleus. As a result, the electrons shell, and thereby, R_{ES}, gets 'stretched'. They make it around, but because of the speed, they have a longer path, and thereby larger R_{ES}. The $(Z + N)$ does not change as observed through multiple experiments of before, during and after.

There are again multiple issues in this calculation:

- Angular momentum, the speed of the electrons, as increasing make the same field strength have a larger radius

- Magnetism created by the particle or whole-system movement may increase the attraction

This is great because we can isolate the change to one half, and the other half, the number of nucleus particles can remain constant. The number of nucleus particles, **N**, does not change. At the end of the interactions causing space-time calculations are a physical known, we still have the same element as each atom. We know the ending atom is exactly the same as the initial atom in number of protons, neutrons, and electrons.

Fundamental Question: What Explains why Mass Changes in Bonding?

Wikipedia says:

"Classically a bound system is at a lower energy level than its unbound constituents, and its mass must be less than the total mass of its unbound constituents."[iv]

What changes is R_{ES}, not (ZN) (the number of nucleus particles).

When you add another atom, and share the electron between them, that structure does not have the protons all in the middle. That changes the R_{ES} calculation. The space between new combo-nucleus is a dead zone, and that is exactly where the bonding electrons sit.

The total # of N nucleus particles does not change, but number of those particles that contribute and how they contribute to the gravity calculation do change. The bonded electrons stop part of their contributing to the gravity formula.

To me, M(NP) is the underlying source of mass, the energy element,

$$m = \int_0^{Z+N} \left(\frac{M(\#_{ZN})}{\frac{8}{3}R_{ES}^3}\right)$$

not which is the 'observed mass'. R_{ES} is a reducer as that volume over which that mass energy got spread at atomic steady state. Only because R_{ES} is a direct relationship to (NP) (but not direct to charge), then the 'observed mass' and 'mass' solve all equations equally as well. Of course, Bohr had to add permeability for that translation. Newton had to the Gravity constant for its translation. Each solution has some magic, but constant factor to implement the ratio of Magnetic Force, M(), with the 'observed mass'

$$m = \int_0^{Z+N} \left(\frac{M(\#_{ZN})}{\frac{8}{3}R_{ES}^3}\right)$$

The extra nucleus in a different centering point changes the average distance of the set of electrons. This factor is smaller, but offsets the bonding lack of contribution.

Changing the average distances creates a slight change in the Charge/distance-square electron-vs-nucleus differential. That changes the force projected to other atoms – the 'mass.'

As such, the differential is slightly different, and we can measure the change in 'mass' which is a change in the average electron distance of the combined molecule versus the atoms operating autonomously.

A picture explains. In an atom, the electrons must be all pushed out relative to the nucleus. That gives a consistent ratio which measures as a steady mass. However, in a molecule, that is bonded atoms, the nucleus sits in different places. As a result, those electrons that are between the nucleus1 and nucleus2 do not contribute to gravity. In fact, they have the reversing effect, the nucleus is outside those electrons so that reduces the atom-gravity.

(55)

Bonding Electron Not Contributing to Gravity / Mass

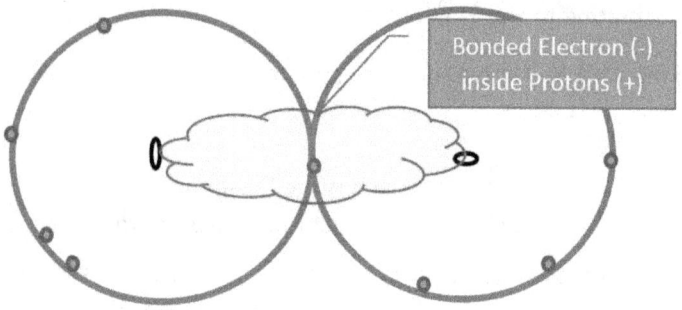

Bonded Electrons do not contribute to the Full External Magnetics (Atom-gravity) 'mass' observed, Instead Push Nucleus Apart

The electron in the bonding position does not contribute to charge-differential (atom-gravity 'mass'). In fact, it decreases the calculation overall. Instead of electrons outside protons consistently, this event is electrons inside protons.

(56)

Bonding – As If Electron Removed From Mass

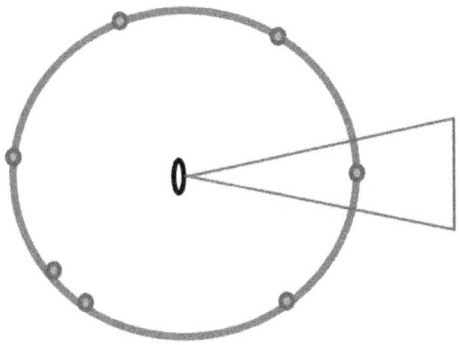

But there are 'reflections' that mitigate that. One great set of mathematics described something similar as the Fresnel zone. In my real job, I have built cellular telecom systems and more specifically microwave towers to connect those towers. We needed to check Fresnel for each tower. Microwaves fail even if the line of sight is clear because of the other 'Fresnel zones.' Between two towers in a narrow window (like the electron-bond cloud), in increments of $1/n$ wavelengths, that Fresnel gets calculated, if there are any objects in that red cloud, the signal gets messed up. That calculation can get messy, but generally, all you really need to know is do not let trees even in the 'curved space' between microwave towers. A tree, wall or parked truck even not directly between can creates reflections that change the base microwave operation.

The slightly different logic in the details applies to atom-gravity 'mass', but it has similar complexity. While the electron does not contribute to charge-differential (atom-gravity 'mass'), there are 'reflections' that mitigate that.

One would expect that the 'mass' lost in bonding would get reduced by that reversal of:

Electrons bonding / Total # of Electrons

However, experimental evidence shows the change is not that dramatic.

Decrease offset in that Electron still contributes to push out other electrons

The observed-mass reduction is one step more complex. While that electron does not create a charge-differential, it does contribute to keep the other electrons out in their orbit which means that the average radius does not decrease.

If you take out that electron, then the other electrons would have been closer in. That electron, even in the bond, still contributes to the electron-shell radius (distance for 1/distance-squared).

Without that bonding electron, the other electrons would have position closer together. Without that electron (red triangle) the other electrons (purple) would not have the electron-electron repulsion (orange arrow) from that missing direction. As a result, the bonding electron still does some contribution to the electron-shell radius.

(57)

Bond Electrons Do Add Part of Mass that Pushes Other Electrons Outside Proton-Proton Vortex

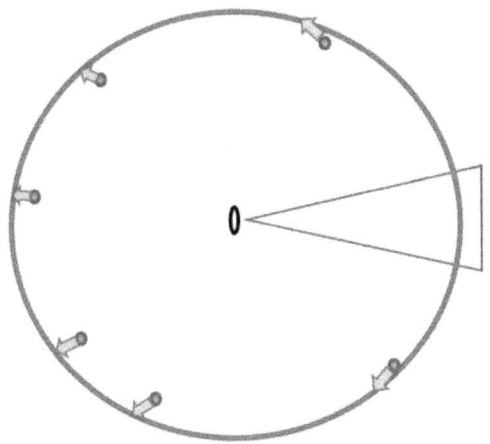

The electrons-in-bonds impact on shell distance can get describe as:

(58)

	Electron-Electron Repulsion	Electron-Electron Repulsion	Change to atom-gravity
In Atom	100%	100%	
When Bonded	100%	(n-1)/n	Pro rata

Decrease offset by increases in distance to the electron in another atom in the bond.

In addition, the bond distance also now goes in part to both nucleus locations. That makes distance (and thereby atom-gravity) increase.

(59)

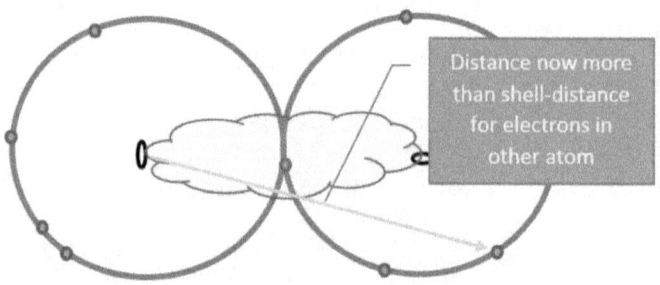

Please notice that in standard bonding, it is only one electron in the space from one nucleus to the other nucleus. The 'paired' electron for each atom is in the opposite hemisphere at the exact opposite position in each atom. AVSC bonding is not the 'Pauli sharing' logic. Pauli's logic gets the same answer for the counts of electrons required, but the Pauli logic does impact incorrectly the force calculation to use the AVSC single-electron.

Remember that everything in AVSC has a 3D picture and longitude/latitude calculation – down to every particle.

Must also realize location is not fixed, but a quantum path

Further, each electron floats, so, in its quantum path, that electrons are outside that zone some percentage of the time and does contribute in those periods.

Calculations done with a perfect circular orbit or exact electron location yield a result up to a factor of 0.6 too small. However, in quantum-path, that electron predicts to be closer and farther away. However, again the 1-distance-squared makes the times when quantum-walk closer have huge impact.

Conclusion: Too much to calculate here

So, I will report in a few years and update this book when all those elements get calculated. I have not done those.

However, in generally, the atom-gravity ('mass') of the combined molecule must go down when compared to the individual atoms. I expect the mass reduction is a Fresnel zone calculation. Those electrons shared contribute less than standard to the charge-differential calculation that is the basic of atom-gravity ('mass').

What is the Einstein's Energy? How do the Units of Measure Work!

How is it that what Einstein was measuring is speed – mass times speed-squared? And that equals energy?

In basic math, the form of the speed of light is as meters per second. Then you should transform that into:

Mass times meters-squared per seconds-squared.

(60)
$$\frac{kg * m^2}{s^2}$$

You move a particular weight in one direction only getting faster by so many seconds every so many seconds; hence kilograms times meters, then divided seconds-squared (seconds per second).

What is the extra meters?

The AVSC Atomic Model shows the replacement of the weight (kilograms)

(61)
$$\frac{Force}{m^3} * \frac{m^2}{s^2} = \frac{Force}{m * s^2}$$

Imagine you have a string of neutrons that is a meter long. How powerful would that magnet be?

(62)

You can observe something close to that at Stanford's SLAC quantum accelerator. They need magnets that big or bigger to direct particle beams.

Those linear accelerators have huge magnets, but even that is not the full strength of Einstein's equation because there is a balance set of electrons spaced out an electron shell radius, then spaced to the next molecule of a stable atom magnet. That means that the above has 1,800x for electrons, and another 3x plus of that for molecular bonding more particles of nucleomagnetics than even the strongest quantum accelerator. The magnets they use has a nucleus then lots of space, then an electron, before it gets to the next. Because the magnetic particles have electrostatic charge, the protons (+), then electrons (-) which substantially offsets its power.

My next proposal for those is just that – see how long a chain of neutrons can get created, and from that how massively powerful is the magnetic field. It is a quantum accelerator that can fit into your living room.

The reality is that we cannot get particle chains that long. The force to make them twist back onto them back into

Further, the AVSC nucleomagnetics is not directional. Only when you rotate particles in loops do you get that north-south of motomagnetics, the magnetics commonly known.

(63)

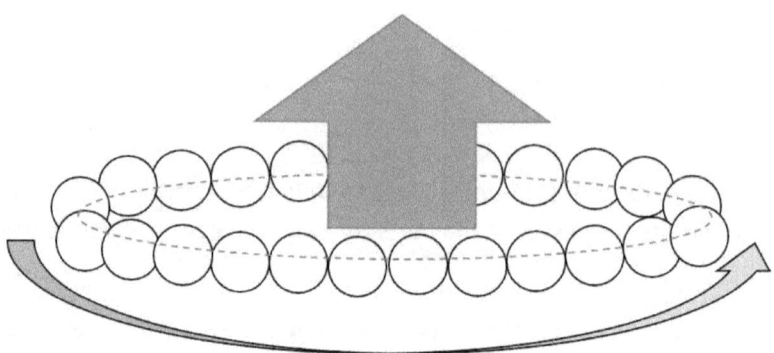

That means that most likely the chain binds into loops, creating a right-hand rule for motomagnetics force. Further, to get the maximum force, let us rotate it at the speed of light. At the distance of one meter, that force is Einstein's formula.

The reality of constructing such a quantum accelerator is challenging because it grabs everything close to it. You would have to accelerate neutrons towards a staging area. In that area, you would need to:

A) Neutrons are kept in a space devote, substantially, of protons.

B) A neutron is accelerating towards the structure until a string of neutrons exist.

C) That is difficult because too much force on the approaching particles will trigger neutron decay.

D) At some point, the chain will flop around until it becomes a ring as the loose ends are highly incented to bind at the particle level. Once in a ring, the system has a high hurdle to unlink back into a chain. It is a one-way path, building most nucleus structures into P-N-P-N rings of different size and count.

E) A few protons spaced can become beneficial. The protons in the ring creates grab points where we can accelerate or decelerate with corresponding electrons flows.

F) The chain once created is keep at high speed by a standard electrical current running outside in the direction of choice. The more speed, then angular momentum holds the shape and structure of the large empty ring.

G) Additional neutrons are accelerated to enter the chain in its rotational direction tangentially. The target changes as the chain grows.

This would become a directional magnetic force. It would also be the largest atom ever created.

That means that the AVSC Energy become force in one direction. Force, the Magnetic Energy, per meter, in one direction which makes sense.

(61)
$$\frac{Force}{m^3} * \frac{m^2}{s^2} = \frac{Force}{m * s^2}$$

Better, when resolved, the base unit is the Magnetic Force; that is a fixed number; it is quantitized; it does not change; and it is not infinite!

Mass is the bundle of energy of the nucleus particles, as constructed, when in a stable atom.

(62)
$$m = \int \frac{M_A(Z_1 + N_1)}{2 * \frac{4}{3}\pi(R_{ES})^3}$$

Redefining the Quantum of Energy

AVSC proposes that there are two basic forces in opposition in the universe. For an electron the nucleus proton, one force is electrostatic charge (blue) and the other is nucleomagnetics repulsion (green). The next of these are repulsive at small distances, but attractive at long distances. The net of those two is balanced to zero at one point (red), the average radius of the electron shell (R_{ES}), effectively the Bohr radius.

(63)

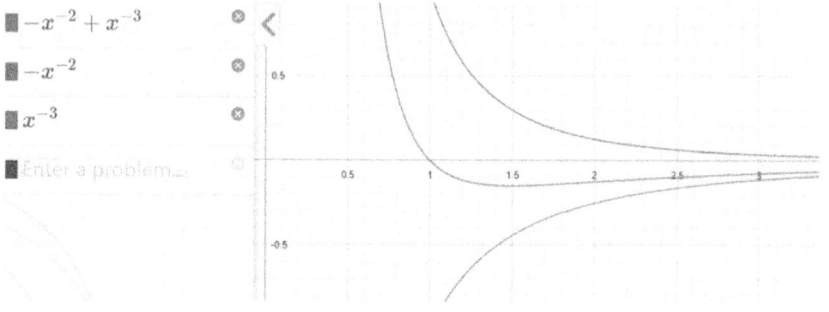

There is one more thing to understand is that the magnetic strength changes by the inclination angle. Bohr got 01-H Hydrogen atoms where the electrons sit on the magnetics axis, so the calculation above is the straight comparison (the red line). However, for electrons in larger subshells will sit off the axis, and at more repulsion.

(64)

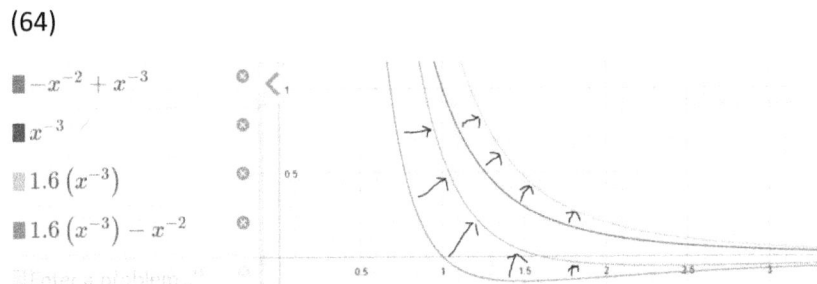

The amazing fact is this big attraction is this space less than zero, it looks huge (yellow area). However, the distance in that area is 10E-11 meter. It is a very tiny space that is repulsive. On the other hand, the attractive area at long distances goes on forever (red). The purple area is the inside of the nucleus particles, which thereby stops the repulsion from going to infinity. It keeps the function continuous in time and in space.

(65)

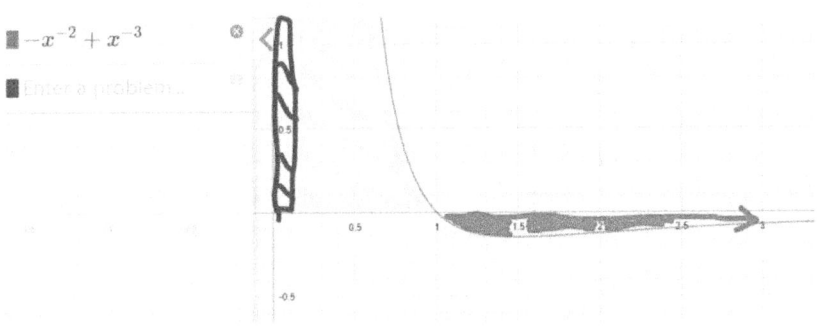

Counter Argument: At zero, the nucleomagnetics become infinitely larger, so how can they balance.

Yes, in fact, strangely, 1-distance-cube is infinitely larger than 1-distance-squared.

But, particles have dimensions. And the electrostatic charge and the nucleomagnetics builds up over the size of that particles. As such, it cannot get to 1/zero = infinity.

However, that balance calculation only works if you exclude the area volume inside the actual particles. This treatise does not address the insides of particles. That is much too complex. The idea is that once the particles have built this complex Newtonian balance.

(65)

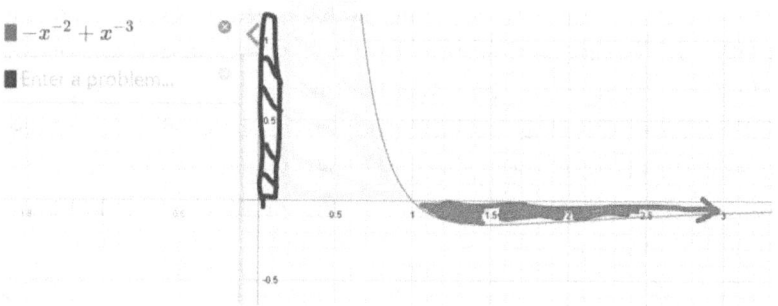

So, there is a big space between the electrons on the axis, and electrons at the equator. This is the next twenty years of my calculations to push this to resolve Schrodinger's Equation with a AVSC function that always resolves (the current Hamiltonian solutions have both converging and diverging solutions which will always converge with the addition of nucleomagnetics force is added appropriately.)

(66)

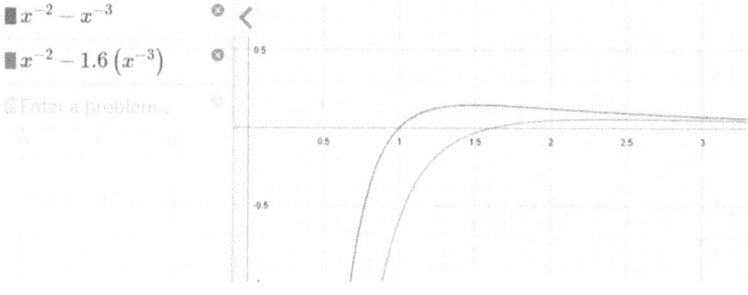

Therefore, the amount of force that exists in a quantum of energy, is the release of one particle, but from that special distance.

The quantum of energy is not the force of a particle, but **the force of a particle after the R_{ES} reduction**.

Now, because of the nature of angular momentum, that calculation comes to the same as Einstein applied. That took them in the wrong direction a century ago.

The quantum calculations were done, and work best, for a Hydrogen atom, where an electron is added, but the nucleus count does not change. It is only half of the nucleomagnetics balance. The next particle adds one side of the multiplication, but not the other. The quantum of energy is **the energy of an added particle to an existing balanced AVSC atomic system (an atom or a molecule)**.

This is best to see in the calculation of the R_{ES} for Helium. In AVSC, an atom of the Element 02-He Helium basic He_4 isotope, is a nucleus ring of 2 protons and two (2) neutrons with 2 electrons. The protons and neutrons assemble in a line or a ring (I will use the more pleasing ring nucleus structure for this example).

(67)

- ●   Proton
- ○   Neutron
- ◎   Electron
- – · –   Nucleomagnetics Axis

However, by calculation the electrons do not sit at the 01-H Hydrogen radium, there is an extra electron which push each other further out.

For Hydrogen, the calculation is simple. There is only one proton, zero neutrons, and one electron. All settle on the nucleomagnetics axis. The calculation becomes:

(68)

Calculation of Bonding Angle & Positions by Element
For the Electron in 001-H Hydrogen - Electron 1m1

Calculation of Forces just for various electrons to the Nucleus

Coulomb's Law	k =	$8.9875517873681764 \times 10^9$ N·m²/C²	
Bohr Radius	r =	$5.2917721067(12) \times 10^{-11}$ m	
Present Invention Constant	m-sub-A =	4.76E-01	

Element		Charge-Force Constant	# Particles	Gross Q-Force	Distance			TF/d2 = Q-Force
001-H Hydrogen	Nucl<>1m1	8.99E+09	-1	-9E+09	5.29E-11			-3.2095E+30
Total	1m1							

		Magnetics	# Particles	Gross M-Force	Distance	TF/d3	COS-Angle	M-Force
		4.76E-01	1	4.76E-01	5.29177E-11	3.21E+30	1	3.20951E+30

Spherical Position	x =	0 y =		0 z =	5.29177E-11	
		Mag.Angle		0		

	Net Nucleus Force	Ratio
	0	-100.00000%
	0	0.00000%

This calculation comes to a net balance of forces at the Bohr radius.

However, the calculation become more complicated with 02-He Helium. Now there are 2 protons, 2 neutrons, and 2 electrons. Further, the electrons are at opposite ends of a structure.

However, the nice thing is all the particles lie on the nucleomagnetics axis. That means that we do not need to add the inclination/longitude adjustment factor.

Therefore, we have attraction (-2) from the electrons-protons, yet repulsions of (-4) from the nucleus total of two (2) protons and two (2) neutrons. As a result, the 02-He electrons sit at a further distance than Hydrogen:

(69)

Calculation of Bonding Angle & Positions by Element
For the Electron in 002-He Helium - Electron 1m1

Coulomb's Law   k = $8.9875517873681764 \times 10^9$ N·m²/C²
Bohr Radius   r = $5.2917721067(12) \times 10^{-11}$ m
Present Invention Constant m-sub-A= 0.475601

Element	Electron(s)	Charge-Force Constant	# Particles	Gross Q-Force	Distance	N/D^3	(1+3COS) factor	TF/d2 = Q-Force
002-He Helium	(NZ)<>1m2	8.99E-09	-2	-17975100000	1.21E-10			-1.22864E+30
002-He Helium	1m1<>1m2	8.99E-09	1	8987550000	2.42E-10			1.5358E+29
Total	1m2							-1.07506E+30

		Magnetic s	# Particles	Gross M-Force	Distance	TF/d3	COS-Angle	M-Force
as above	(NZ)<>1m2	4.76E-01	4	1.902402653	1.20955E-10	1.08E-30	1"	1.07506E-30
as above	1m1<>1m2	4.76E-01	0	0	2.4191E-10	0.00E+00	1"	0
Location	1m2							1.07506E-30

Spherical Position		x =	0 y =		0 z =	1.20955E-10		
			Direction					

as above	(NZ)<>1m2							-1.07506E-30
as above	1m1<>1m2							1.07506E-30
								6.71909E-18 0.00000%

This comes within 2% of the established standard electron position for 02-He 1s2 (1m2 in AVSC) electrons. That difference is that I have not included in this calculation: a) the angular momentum rotating the electrons (as the above assumes zero Kelvin); and b) the loss of magnetic force by the proton-neutron nucleus structure (it is not a point as assumed in this simplified calculation), and c) any quantum mechanics for the systems own harmonics. Of course, d) the inclination/longitude force amplification part of the AVSC calculation was also not required for 02-He.

Calculation of Quantum of Energy is the Adding of One Electron to the System

In this example, imagine that we would ionize a 02-He. I know that is not particularly stable, as 02-He Helium atoms are a full-shell noble gas. That extra electron will not push all the other electrons, but not change the nucleomagnetics the base calculation.

(70)

 New Particle

- ● Proton
- ○ Neutron
- ◉ Electron
- – · – Nucleomagnetics Axis

Every settling position gets adjusted by the secondary energy of that particle added to the system. Of course, that moves those particles, and those electrons (particles) then eject a Newtonian balancing electromagnetic wave.

The rest of the chain has already been proven by others. The relationship of the Planck constant to the electrostatic charge to the speed of light.

Planck constant is 6.626E-34 Joule-seconds or 4.135E-15 eV-seconds.

The electrostatic charge is 1.602E-19 coulomb which is also one (1) eV.

(71)
$$1\frac{J}{C}\sqrt{\frac{2\hbar^2}{\mu c}}$$

The momentum of each electron getting an additional electron into the system will have its momentum changed by the quantum of energy, the reflection of the added particle pushing the existing particle but only as reduced by the R_{ES} factor.

In this sense, this postulate moves away from the notion that the quantum of energy is its own particle.

Reflection of New Particle Creates Other Particle Changes, So Field Change Electromagnetic Waves

However, that itself does not resolve fully why these creates electromagnetic waves. There are waves created by these interactions. The disruption of the nucleomagnetics field positions and orientation is the traditional electromagnetic waves, the motomagnetics of changes to standing nucleomagnetics fields and patterns.

In a separate postulate, I will review how the additional particle changes the positions and rotation of existing particles. Thereby, the nucleomagnetics field of each changes in period rotations (ah! Magnetics rotating makes electromagnetic waves). As such, we can have waves of magnetics from that change. However, that concept itself will take an entire book to detail.

Remember that particles are rotating (heat) such that they have a change of inclination, and nucleomagnetics force going on all the time before the new particle. The new particle changes that in distance, in orientation, in angle, and thereby changes the standing motomagnetics field in these specific jumps.

The changes to the electrostatic field do not creates these waves. That is because the wave is spherical, so there is no whipsaw. Rotation does not cause a standing electrostatic field. So, the new additional particle does not create a pattern. Electrostatic fields are stable and continuous.

Rotation does create a cyclical magnetic wave – and a frequency – which is the basis of 'electromagnetic' waves. That is the changing field as the particle rotates.

Challenge: What Keeps Electrons from Falling into the Nucleus?

Charge is the Most Powerful Force in the Universe

Charge-force lights our cities so bright that anyone can see it, as electricity, at work from deep outer space.

(72)

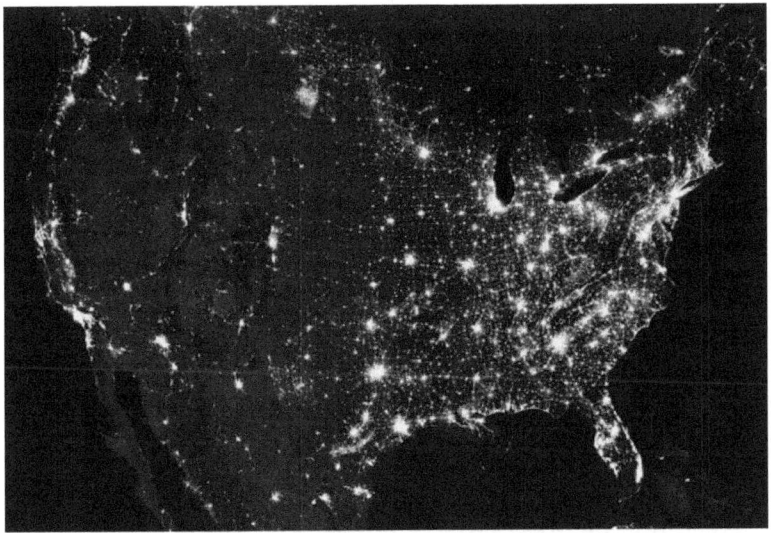

Protons, with their positive electric charges, are powerful. And protons positive (+) charges attract the negative charges of electrons, yet electrons in billions and trillions of atoms do not fall into the nucleus protons. This is a fundamental question of the nature of the world.

Protons in the Nucleus Strongly Attract Electrons

Charge-force follows the rule:

- Like-Charges Repel
  - Positive (+) to positive (+) repels
  - Negative (-) to negative (-) repels

- Opposite-Charges Attract
  - Positive (+) to negative (-) attracts
  - Negative (-) to positive (+) attracts

Keeping electrons in the electron shell is very challenging. Under normal circumstances, the charge force would bring the electrons and protons crashing together.

The answer lies in nucleomagnetics, the subatomic particle force that creates our big-world, traditional magnetics.

1) But the electrostatic charge strength is substantially greater - between 10-100 times stronger at the one meter apart.

(73)

Factor for 001-H <> 001-H atoms@1m	Force Calculation	
Electrical Charge Force	10^{-28} m^1 / (s^2)	*
Magnetic Force	10^{-30} m^1 / (s^2)	*

Solving how the strongest force in the universe does not make atoms collapse into a nucleus is the adventure. The basic question is: **Why does an electron stay away from the nucleus if charge-force is so strong?**

Describing the Forces

The two strongest forces, charge-force and magnetic-force, work in balance within a nucleus, but before I describe that relationship, I will review how charge-force and magnetic-force are the same and how the two differ.

Charge-Force is Spherical

Charge force goes in every direction. At any distance, in any direction, the charge is the same. This applies in 3D, the X, Y, or Z direction.

(74)

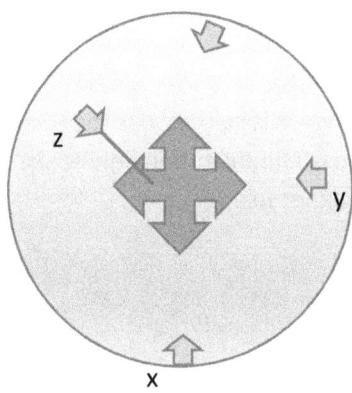

At the same distance, the field strength (yellow arrows) is the same.

Magnetic Force is North-South Oriented

Magnetic-force is north-south oriented. It is strong at 90 degrees and very weak at the poles.

(75)

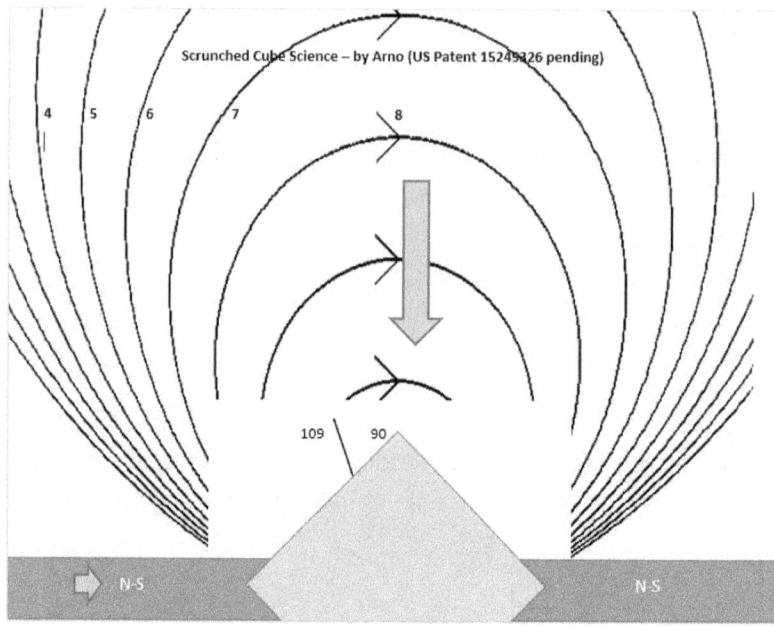

Nucleomagnetics-force is stronger at the 90 degrees, and weaker at the magnetic poles. The magnetic field is like a 'bagel', bulging at 90 degrees to the magnet itself.

Charge-Force decreases by 1/distance-squared, Magnetic Field Strength decreases 1/distance-cubed.

The strength of an electrostatic charge force comes from the proton (+) or the electron (-). Once that charge exists, its strength decreases based upon the volume.

If you have an electrostatic charge at a small area 1x1x1 as a starting point. For our calculations that is:

(76)

$$\text{Strength} = \frac{2(\sqrt{k})(Q)}{(1)^3}$$

$$\text{Force} = \frac{(\sqrt{k})(Q)}{(1)^2}$$

For an example twice, as part away, then:

(77)

$$\text{Strength} = \frac{2(\sqrt{k})(Q)}{(2)^3} = \frac{2(\sqrt{k})(Q)}{8}$$

$$\text{Force} = \frac{(\sqrt{k})(Q)}{(2)^2} = \frac{(\sqrt{k})(Q)}{4}$$

The makes sense. If you have a strength of 8 at a point, then distribute that energy to a field that is 2x2x2, then the volume of 8 which makes each volume, now spread 1 in strength, with 8 boxes.

At the initial box, the strength is 8.

(78)

   1 box x 8 strength = 8 total field strength

When distributed over 8 of the same volumes, then the strength is 1 for each box (1/8th of the initial strength). That way the same total energy is conserved. Eight (8) boxes of 1 is the same as one box worth eight (8).

(79)

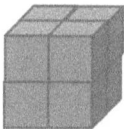 8 boxes x 1 strength = 8 total field strength

Now the real answer is a sphere, with $\frac{4}{3}\pi r^3$, but the picture is hopefully enough to show the concept that a charge distributes evenly across the volume as it (the field strength) spreads to a distance.

So, for a position at radius two, that field has the same strength, but it is distributed over the entire volume. Therefore, at 2x (twice) the distance, the field is 8x stronger ($\frac{1}{d^3}$).

(80)

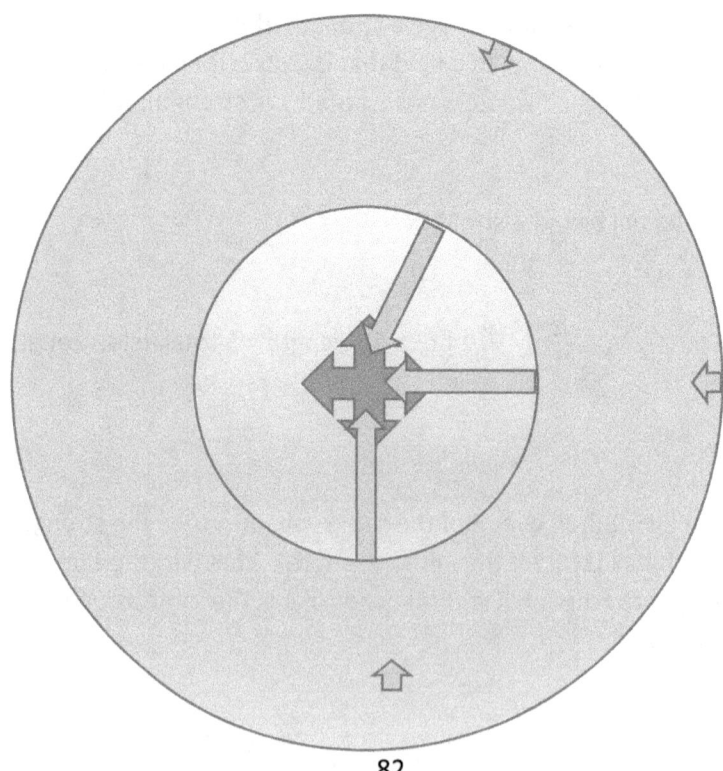

Yet, if you are looking at this from one direction, then that one strength factor does not decrease, you get all the strengths in that direction. Therefore, the force decreases by 1/distance-squared ($\frac{1}{d^2}$)

(81)

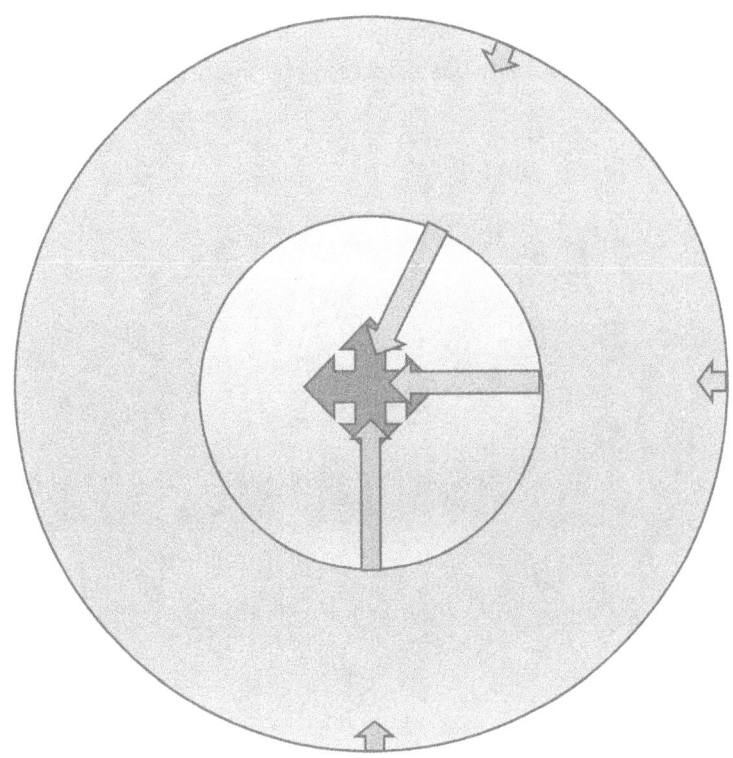

Think of it this way. In the direction of the distant object, all those strengths (green arrows) are still applying fully. So, in one dimension, the field strength does not decrease. That leaves only two (2) dimensions (a factor with base $\frac{1}{d^2}$) that decrease with distance **_relative to_** a distance object in one direction.

Magnetic-Force decreases by 1/distance-cubed plus angle versus north-south, Force decreases by 1/distance-squared at north-south and 1/distance-cubed at 90 degrees.

A magnetic field already has a decrease by the oriented direction (purple arrows).

(82)

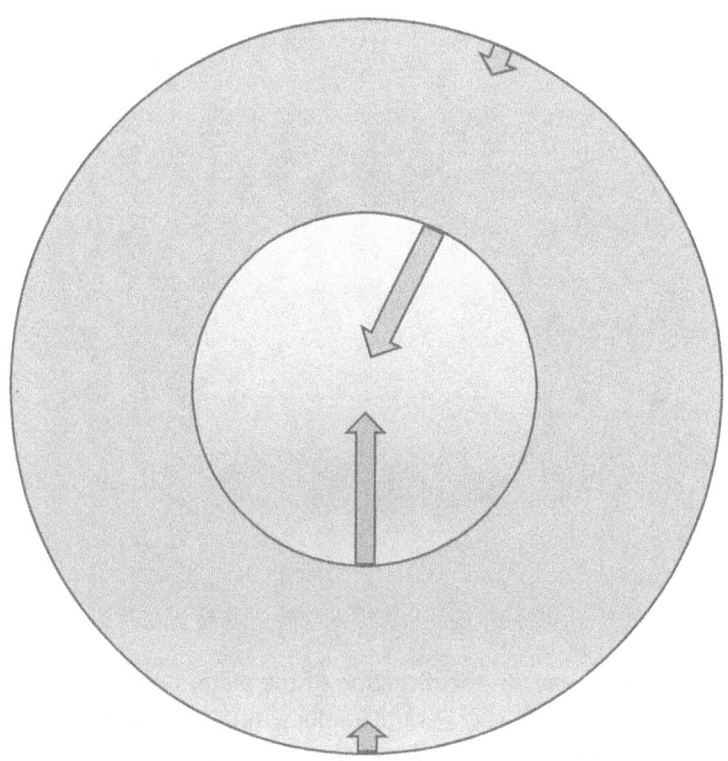

Therefore, from that direction is a) already low, b) decreases by $\frac{1}{d^2}$. Yet, in the other directions, the magnetic force gets pulled down by orientation dimensions, so $\frac{1}{d^2}$ becomes $\frac{1}{d^3}$ based upon the angle after the end of the magnet. Or you can think of that as $\frac{1}{(d^2)(d*\arcsin\ )}$

Charge-Force begins the reduction at the edge of the particle.

The magnetic field starts decreasing that edge of the particle, and for out purposes, that is a tiny sphere in the middle of the field.

(83)

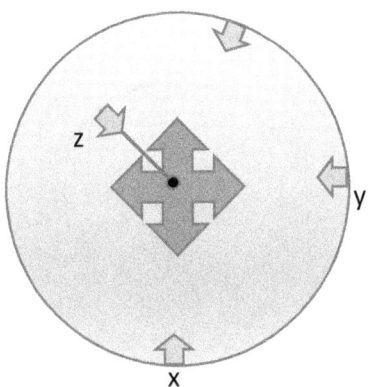

Charge-force decreases in every direction, and thereby, all of the field vectors point to the center of the origin particle.

The center of a charge-particle is also the center of the charge-force. Therefore, even if the particle has dimensions (X, Y, Z), that charge-force centers in the same place as center of the particle.

Yet, that is not the case for magnetic forces.

Magnetic-Force begins the reduction at each of the poles.

The magnetic field remains the same along the width of the magnet. It does not decrease its strength until it gets to the end of the magnet.

(84)

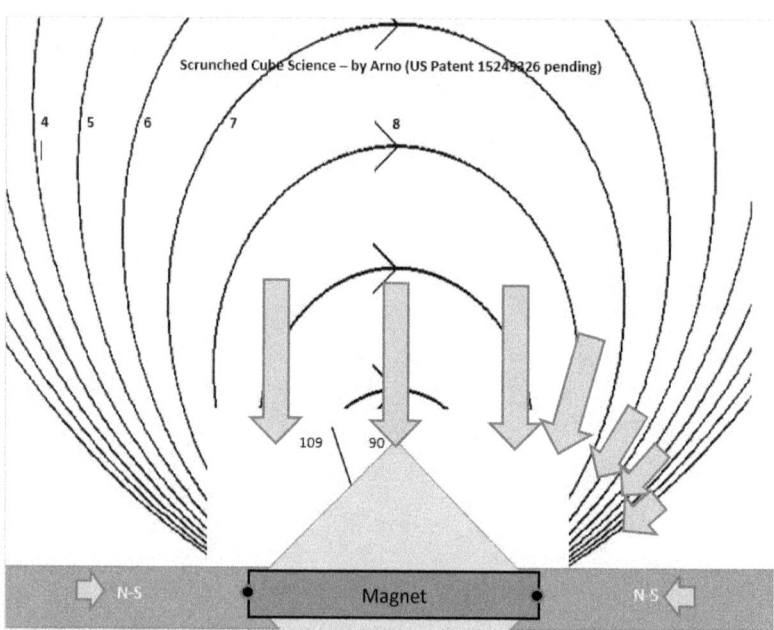

Further, the field points toward the nearest pole, not the actual center of the magnet. The field vectors point to a place different that with charge-force.

The magnetic field really points to each of the ends of the magnet when past that end. In the middle, they point to perpendicular – toward the nearest point on the magnetic bar.

In this tiny way, charge and magnet-force are different.

Summary:

Therefore, in terms of initial strength, Charge-force is stronger:

(85)

Factor for 001-H <> 001-H atoms@1m	Force Calculation
Electrical Charge Force	10^{-28} m^1 / (s^2) [vi]
Magnetic Force	10^{-30} m^1 / (s^2)

Therefore, in terms of rate of decrease, Charge-force is again stronger:

(86)

	Force Decrease Factor
Electrical Charge Force	$\frac{1}{d^2}$
Nucleomagnetics Force factor	$\frac{1}{d^3}$ at the nucleomagnetics poles $\frac{\sqrt{2}}{d^3}$ at the nucleomagnetics equator And $\frac{\sqrt{(1+3COS^2\theta)}}{d^3}$ between

First, at this level, magnetism totally rules. Even if a charge got closer, the magnetic force would still be stronger.

(87)

$$F = k \frac{Q_1 Q_2}{d^2}$$

(88)

Factor	Charge-Force Estimation
Charge-Force Constant (Coulomb)	$k_e = 10^{10}$ [8.99x10⁹ m³ kg²/ (s²)]
Charge Proton	$q_1 = 10^{-15}$ [0.9×10⁻¹⁵]
Charge Proton	$q_2 = 10^{-15}$ [0.9×10⁻¹⁵]
Distance	d=3.0 x 10⁻¹⁶ m or 10⁻¹⁶ m
Exponent shortcut	+k+Q+Q-d-d
Short-cut calculation	10-16-16-(-10)-(-10)= 10-16-16+15+15= 10-32+30=8 N (Newtons)
Charge-Force Repulsion	10⁺⁸ m¹ / (s²) (Newtons)

(89)

$$F = M_A \frac{(NP_1)(NP_2)}{d^3}$$

Factors	Magnetic-Force Estimation
Charge-Force Constant (Coulomb)	$M_A = 10^{-38}$ m³ kg²/ (s²)]
Charge 001-H Hydrogen atom	n=1 or 10⁰
Charge 001-H Hydrogen atom	n=1 or 10⁰
Distance	d=0.9 x 10⁻¹⁵ m or 10⁻¹⁵ m
Exponent shortcut	+k+Q+Q-d-d-d
Short-cut calculation	-38+0+0-(-15)-(-15)-(-15)= -38+0+0+15+15+15= -38+45=+7 N (Newtons)
Magnetic-Force Attraction	10⁺⁷ m¹ / (s²) (Newtons)

Yet, if we keep a neutron between the two protons (+), then the calculation becomes the magnetic force is stronger - voila:

(90)

Factor	Charge-Force Estimation
Charge-Force Constant (Coulomb)	$k_e = 10^{10}$ [8.99×10^9 m³ kg²/ (s²)]
Charge 001-H Hydrogen atom = 1 proton	$q_1 = 10^{-19}$ [1.67×10^{-19}]
Charge 001-H Hydrogen atom = 1 proton	$q_2 = 10^{-19}$ [1.67×10^{-19}]
Distance	d=0.86 x 10^{-15} m or 10^{-15} m
Exponent shortcut	+k+Q+Q-d-d
Short-cut calculation	10-19-19-(-15)-(-15)=
	10-19-19+15+15=
	10-38+30=+2 N (Newtons)
Charge-Force Repulsion	10^{+2} m¹ / (s²) (Newtons)

(91)

Factors	Magnetic-Force Estimation
Charge-Force Constant (Coulomb)	$M_A = 10^{-38}$ m³ kg²/ (s²)]
Charge 001-H Hydrogen atom	n=1 or 10^0
Charge 001-H Hydrogen atom	n=1 or 10^0
Distance	d=0.9 x 10^{-15} m or 10^{-15} m
Exponent shortcut	+k+Q+Q-d-d-d
Short-cut calculation	-38+0+0-(-15)-(-15)-(-15)=
	-38+0+0+15+15+15=
	-38+45=+7 N (Newtons)
Magnetic-Force Attraction	10^{+7} m¹ / (s²) (Newtons)

Even at the nucleus level, the AVSC model of proton 1/distance-square electrostatic charge repulsion and ASVC nucleomagnetics resolve the strong force calculation without additional particles and without discontinuities.

Newton Discovered Both Gravity and Integrals, and this Extension of his Gravity Formula Applies Both of Newton's Discoveries

It is amazing and funny that Sir Isaac Newton was both discoverer of gravity and inventor of the calculus of integrals, but Newton never had to apply the rule of integrals to his calculation of the force of gravity. (Electrons were not even observed and documented until Rutherford, Bohr, and, of course, Faraday a hundred years later. I trust Newton would have figured it out this discovery immediately.)

A gravity-orbit only has the one force with both objects locked into each other, so the integral gets the same result; it is easy to average. Scientist for a century have probably missed the connection.

> *Why Missed?*
>
> When the two gravitational-orbit around each other, the integrals are not very helpful for two reasons.
>
> First, the mass, and thereby force, at each end is homogeneous. As such, there is not that huge different of (+) at the center vs the (-) distributed at the shell radius in the integral. When the shell is (-) that difference become important and measurable. For a planet, the whole distant object is homogeneous. For a nucleus-shell structure, the force is not homogeneous; the charge has opposite signs, positive at the center and negative over the shell surface area. Even with that, this calculation generates only a tiny difference ratio, 10^{-37}, so most people would not make the connection.
>
> Second, the integral for homogenous masses/forces gets back to an ellipse orbit because the two items are the only two interacting. However, for electron orbits and long-distance

> objects, there is more than just the gravity interaction, so the extra effort for integral finds the calculation. The shells create a 3rd factor that is oriented north-south, so it is not easy to measure.
>
> The gravity-orbit mass-at-a-point-distant methods works well enough. Only NASA needs to worry about some tiny variances that would make a spaceship miss Juniper with just is miscalculation of 10^{-11}. By the way, NASA must deal with the homogeneous much smaller distribution of gravity objects at $<10^{-10}$ accuracy or they miss the planet.

In the end, all I really do in this proof is change the long-established Newton gravity formula:

(92)

$$F = G\frac{m_1 m_2}{r^2}, \qquad G = 6.67 \times 10^{-11}$$

This calculation is easy because it assumes that all the mass sits at a point at a certain distance (r) from the distant location. There is no integral; that makes the calculation easy, but misses that fine detail.

For a charge distributed over 1-unit block at distance d, the Newton Gravity formula changes into an integral of the strength field:

(93)

$$F = \int_{x=r-1\,Planck}^{x=r} -2G_N \frac{m_1 m_2}{x^3} + C$$

Let's look at the elements for common sense.

The negative sign makes sense. As the force goes away, the force of gravity goes down. The total force is positive, but that force decreases.

The $1/r^3$ also now makes much more sense. The force spreads out evenly in three dimensions (X, Y, Z), and therefore, it reduces in exactly in the same three dimensions. If you take a sphere of radius 2, the volume of energy enclosed is $A = \pi r^3 = \pi 2^3 = 8\pi$, so the force at any point is $= 1/8\pi$. If that radius expands to 4, then the volume expands similarly $A = \pi r^3 = \pi 4^3 = 64\pi$, so the force at any point is $= 1/64\pi$. The total energy both remain 1 $(1/8\pi)(8\pi) = 1$ and $(1/64\pi)(64\pi) = 1$

The $1/d^2$ force in the Newton equation now makes sense. Only two of the dimensions get weaker relative to a direction. The X direction force adds up, so that direction all the pieces are still pulling the distant object. It really is d/d^3 view which becomes $1/r^2$. It has points along the line so only the ones in two (2) dimensions are not pulling the distant object.

The two formulas are the same with just the addition step of doing the calculation for multiple points. The above formula assumes that all the charge sits at a cube with a diameter of 1.

It is as a bunch of little Newton gravity calculations for every combination of particles:

- Electron-to-electron,
- Proton-to-proton,
- Electron-to-proton, and
- Proton-to-electron.

I love Newton. This formula is actually moving back to Newton, which many current alternatives seem to question by adding non-Newton dimensions to get their results that work. As you review, you will see that using this definition, a number of 'extra dimensions' become a knowable force calculated from the Constant R_{es} (Radius-Electron-Shell) and the G_A or $\sqrt{G_A}$ knowing the secret that $\sqrt{G_A}$ actually equals the fundamental particle magnetism.

Let me be clear. Newton was 100% correct. My postulate is a method to apply *Newton* gravity as a *Newton* integral.

Magnetics of Nucleus Particles in Chains Causes Enough Magnetic Attraction to Balance Charge Repulsion

In the rest of this proof, we will move between applying a) the rules that apply to, and b) the outcomes that result from, these two most-powerful forces as we build the basic structure of a nucleus, and then the nuclei of different sizes, shapes, and elements within the periodic chart and their isotopes.

First, the problem is strength. Charge for is stronger.

(94)

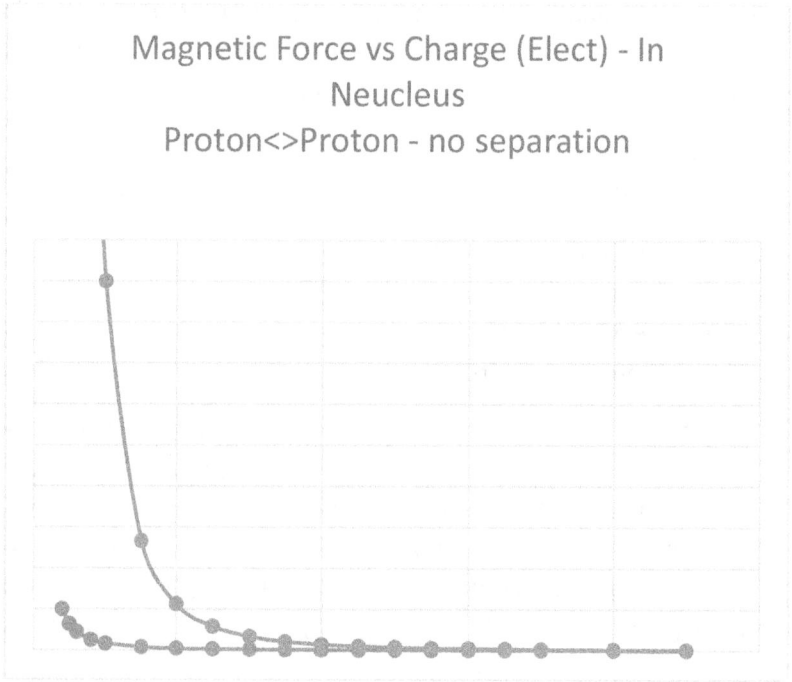

> For the Advanced Explorer:
>
> If particles have no dimensions, as some contend, there is challenge is at very close the 1/distance-cubed would become infinitely huge, but I disregard as the particle must have dimensions and thereby dimensions less than a particle size where that anything divided by zero (1÷0) = **infinity** would not and does not apply. Zero dimension charge particles (protons, electrons) do not exist.
>
> Therefore, this proof assumes that particles do have physical dimensions (length, width, height).

The repulsive force of same-charge between two protons diminishes based upon the square of the distance ($\frac{(\sqrt{k})(Q)}{d^2}$). The attractive force of magnetism diminishes based upon the cube of the distance ($\frac{M}{d^3}$) depending upon orientation.

If you throw protons fast enough to penetrate a nucleus, and thereby touch proton-to-proton, you create a nuclear reaction. That repulsive proton-proton force created is explosive. The 1/distance-squared at small distances nearing ($\frac{1}{0}$) become exponentially devastating.

If there was nothing else, a nucleus would not stay together. The protons would push each other away. Then why does a nucleus stay together?

The above looks at those two curves of distance versus force shows that the Electrical charge (repulsion in a nucleus) in red/orange always exceeds the magnetic force in blue if the two forces are at the same place.

But, there is an amazing additional fact:

A Magnet Stays Strong if in a Chain

We already discussed that magnetic do not start decreasing until the end of the magnet. That the magnetic-force points to the poles, not the center of the magnet.

This applies when particles bind N-S-N-S. The new combination is a full magnet, and again the field decrease does not start until the end of the 'combined magnetic structure' at the ends

The magnetic force extends if physically connected. It does not start decreasing until the end of the magnet. That is, a chain of magnets keeps the force at the same strength – while the proton charge repulsion keeps decreasing. When separated by a neutron, you chain protons, so that the magnetic field decrease starts further to the right; the blue line stays flat for that distance before it starts its decrease.

Let's see how that works. Remember that this movement of the charge strength to additional distance only occurs if physically connected. In this case, that connection is a proton-neutron-proton magnetic chain. I move the magnetic-force (blue) line in the chart.

(95)

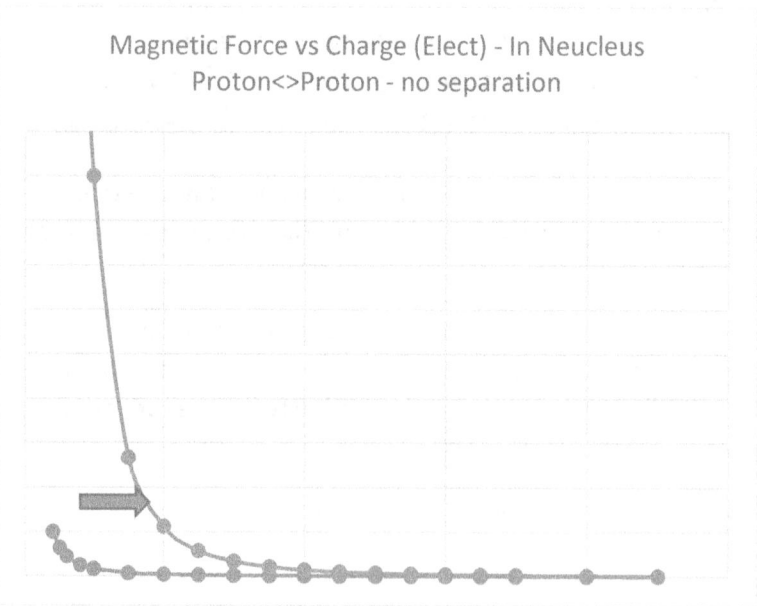

Magnetic Force vs Charge (Elect) - In Neucleus
Proton<>Proton - no separation

This shows that the magnetic force and charge force sometimes trade places. When the charge is moved away, but the magnet-in-chain/still intact remains.

At very short distance, both charge and magnet become huge. However, there is a limit of the physical thing. A magnet builds in the object so someplace at the left, the curve ends. It cannot become infinite because the actual magnet structure is not zero in length. These curves are not applicable at less than the actual size of the particles involved.

There is a place in the middle where the magnetic force is stronger than the charge force.

That new curve with both forces looks like below and for a range the magnetic (orange) force stays above the charge (blue) force:

(96)

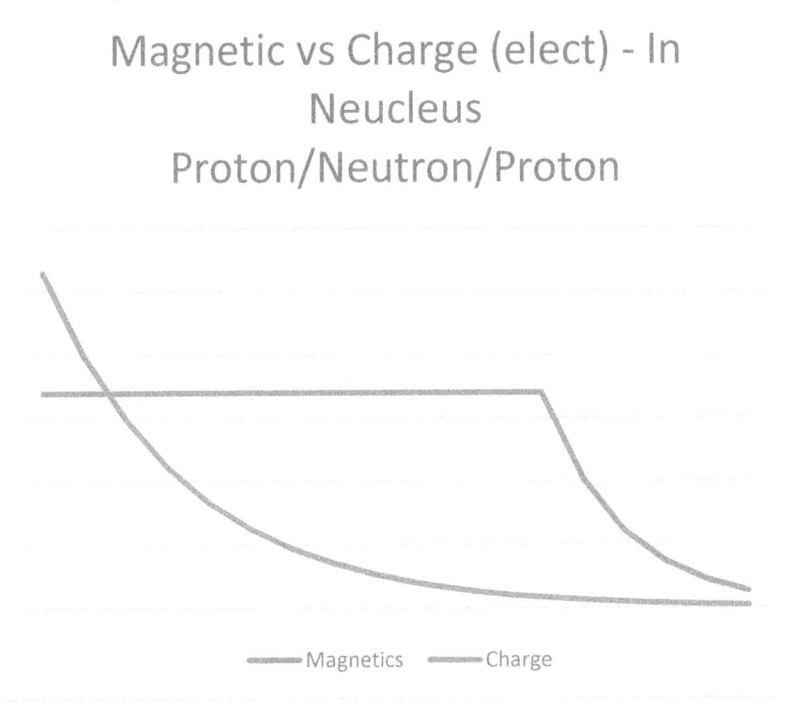

The (orange line) magnetic force I show as constant along a continue chain of magnetic entities which ends at the length of the proton or neutron (about .84 to .87 x 10^{-11}). After that, the curve decreases by the $F = \frac{M}{(d^2)(d*\arcsin\ )}$ such that eventually the charge repulsion is greater at some distance from the nucleus.

The (blue line) Charge force is distance-squared and larger, but only at location closer than the neutron separator.

Now, the above graph shows the two forces, but the charge is repelling and the magnetic is attracting. So, really you need to look at them as the net. That is the size of that different (the space between the two functions which is sometime positive (net repulsion) and sometime negative (net attractions).

(97)

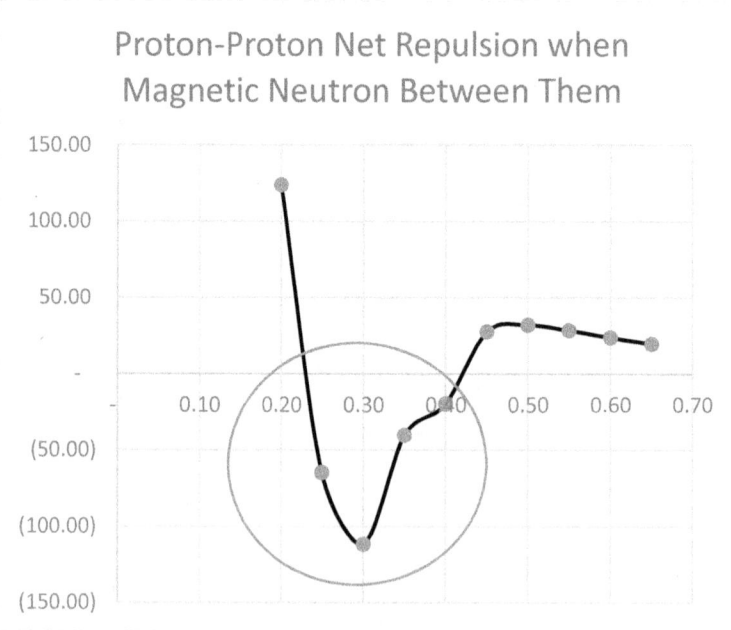

At too close, the structure repels (explosive nuclear reaction force). At far away, the structure also repels (magnetic becomes weaker, faster by 1/distance-cubed). However, at a magic distance in the middle, the structure actually attracts (the area in the middle below zero).

> **For the Advanced explorer:**
>
> For this balancing at long distances, there is a more detailed analysis in my <u>Gravity is Just . . .</u> book. At long distance, the magnetic force is 1/distance cubed (M/d^3) depending on orientation so it becomes smaller than 1/distance-squared (C/d^2). At long distances, only charge is important. The magnetic-force decreases faster, so magnetic force only applies close to the atom.

Magnetic Neutrons are the Required Separators Extending Magnetics While Charge Decreases

What is needed then is a magic particle. A particle that:

- Has no charge

- Is Magnetic strength and orientation

That particle exists; it is a neutron.

With a neutron, atoms get created that have the basic need for building a nucleus structure that is stable. It is a structure proton-neutron-proton and so on . . .

(98)

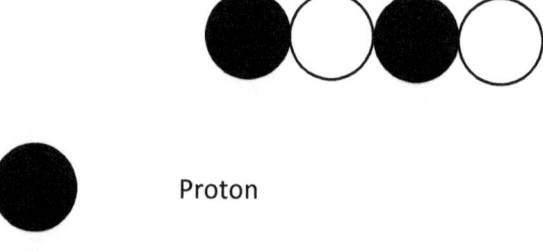

Proton

Neutron

So long as the structure builds from this logic, then a stable nucleus can exist. The distance of a neutron separates enough to lower the charge-force, AND the physical connection maintains the magnetic-force of the chain of nucleus particles.

Calculation of the Strength of Charge and Strength of Magnetic Field at Various Distances of the Radius of a Neutron

There are three basic factors as work in this nucleus binding, the 'strong force' per most textbooks. The proton-proton repulsion, the oriented magnetic field (attractive when north to south), and the physical dimensions of the particles.

1.

Proton-proton repulsion is determined using Coulomb's Law, which is generally understood as decreasing by 1/distance-squared ($\frac{1}{d^2}$).

(99)
$$F = k_e \frac{q_1 q_2}{d^2}$$

2.

The oriented magnetic field (attractive when north to south) is the most difficult, but it is generally understood as decreasing by 1/distance-cubed ($\frac{1}{d^3}$).

(100)
$$F = M_A \frac{n_1 n_2}{d^3}$$

3.

And, the physical dimensions of the particles. Which are actually limits to the distance (d^2). The nucleus particles (protons or neutrons) are generally understood to have dimensions of:

Charge radius   0.8751(61) fm[1] (0.8751 x 10^{-15} m)[vii]

Calculation of Magnetism by Comparison to Charge at Bohr Radius

As noted in Book 1 of Simple Words to Understand . . . Physics Gravity is Just That . . . Electrons are a Little Closer, charge-force and magnetic-force are intricately linked.

At approximately the Bohr radius, the balancing of those two creates the placement of the electron shell. At that distance (d^2), the attraction electron-proton charge-force equals the electron-magnetic field of nucleus repulsion.

The Bohr radius is:

Bohr radius · a_0
$5.2917721092171717 \times 10^{-11}$m r = n^2 a_0/Z [viii]

So,

$$F = k_e \frac{q_1 q_2}{d^2}$$

(101)

Factor	Charge-Force Estimation
Charge-Force Constant (Coulomb)	$k_e = 10^{10}$ [8.99×10^9 m³ kg²/ (s²)]
Charge 001-H Hydrogen atom = 1 proton	$q_1 = 10^{-19}$ [1.67×10^{-19}]
Charge 001-H Hydrogen atom = 1 proton	$q_2 = 10^{-19}$ [1.67×10^{-19}]
Distance	d=5.29×10^{-11} m or 10^{-10} m ($d^2 = 5.29 \times 10^{-11}$)² m = 10^{-21} m
Exponent shortcut *Short-cut calculation*	+k+Q+Q-d-d 10-19-19-(-21)= 10-19-19+21= 10-38+21=-7
Charge-Force Repulsion	10^{-7} m¹ / (s²) (Newtons)

Yet that must be equal to the magnetic repulsion at that same distance.

(102)

$$F = M_A \frac{n_1 n_2}{d^3}$$

Factors	Magnetic-Force Estimation
Charge-Force Constant (Coulomb)	$M_A = 10^{-38}$ m³ kg²/ (s²)]
Charge 001-H Hydrogen atom = one (1) Proton	n=1 or 10^0
Charge 001-H Hydrogen atom = one (1) Proton	n=1 or 10^0
Distance	d=5.29 x 10^{-11} m or 10^{-10} m
	(d^3=5.29x10^{-11})³m =10^{-31} m
Exponent shortcut	+k+Q+Q-d-d-d
Short-cut calculation	-38+0+0-(-31)=
	-38+0+0+31=
	-38+0+31=-7 N (Newtons)
Charge-Force Repulsion	10^{-7} m¹ / (s²) **(Newtons)**

Measuring Charge-Force versus Magnetic-Force at Distance in Nucleus

For the postulate to work, at the levels of the nucleus particles, the magnetic-force must exceed the charge-force.

Of course, given the two are equal at the Bohr radius, and the magnetism will inverse at cube versus the square of charge-force, one would expect that magnetism is must stronger at the distance that of nucleus particles.

That distance would be 2 x radius since it must transfer the one particles and attach to the 2nd.

Let's test this for proton-proton charge-force repulsion at that separation and then for proton-neutron magnetic-force attraction at that separation.

The charge force at a nucleus distance would be: (103)

$$F = k_e \frac{q_1 q_2}{d^2}$$

Factor	Charge-Force Estimation
Charge-Force Constant (Coulomb)	$k_e = 10^{10}$ [8.99×10^9 m³ kg²/ (s²)]
Charge 001-H Hydrogen atom = 1 proton	$q_1 = 10^{-19}$ [1.67×10^{-19}]
Charge 001-H Hydrogen atom = 1 proton	$q_2 = 10^{-19}$ [1.67×10^{-19}]
Distance	$d = 0.86 \times 10^{-15}$ m or 10^{-15} m
Exponent shortcut	+k+Q+Q-d-d
Short-cut calculation	10-19-19-(-15)-(-15)=
	10-19-19+15+15=
	10-38+30=+2 N (Newtons)
Charge-Force Repulsion	10^{+2} m¹ / (s²) (Newtons)

Yet that must be equal to the magnetic repulsion at that same distance. (104)

$$F = M_A \frac{n_1 n_2}{d^3}$$

Factors	Magnetic-Force Estimation
Charge-Force Constant (Coulomb)	$M_A = 10^{-38}$ m³ kg²/ (s²)]
Charge 001-H Hydrogen atom	n=1 or 10^0
Charge 001-H Hydrogen atom	n=1 or 10^0
Distance	$d = 0.9 \times 10^{-15}$ m or 10^{-15} m
Exponent shortcut	+k+Q+Q-d-d-d
Short-cut calculation	-38+0+0-(-15)-(-15)-(-15)=
	-38+0+0+15+15+15=
	-38+45=+7 N (Newtons)
Magnetic-Force Attraction	10^{+7} m¹ / (s²) (Newtons)

So, magnetic-force is huge compared to charge-force at the same nucleus particle distance.

However, the real comparison is that charge-force, if proton-near-proton is much less.

(105)

$$F = k_e \frac{q_1 q_2}{d^2}$$

Factor	Charge-Force Estimation
Charge-Force Constant (Coulomb)	$k_e = 10^{10}$ [8.99×10^9 m³ kg²/ (s²)]
Charge Proton	$q_1 = 10^{-15}$ [0.9×10^{-15}]
Charge Proton	$q_2 = 10^{-15}$ [0.9×10^{-15}]
Distance	$d = 3.0 \times 10^{-16}$ m or 10^{-16} m
Exponent shortcut	+k+Q+Q-d-d
Short-cut calculation	10-16-16-(-10)-(-10)= 10-16-16+15+15= 10-32+30=8 N (Newtons)
Charge-Force Repulsion	10^{+8} m¹ / (s²) (Newtons)

If there is not the neutron separation, then the charge force can get too close, and become 10^{+8} (~10,000,000) m¹ / (s²) vs the magnetic attraction of 10^{+7} (~1,000,000) m¹ / (s²) of proton-proton.

If there is the neutron separation, then the charge force is overcome because the 10^{+2} (~100) m¹ / (s²) charge-force when separated by a neutron is then less than the magnetic attraction of 10^{+7} m¹ / (s²) of proton-proton.

As such, the forces at play do not bond proton-proton, but do bond proton-neutron-proton.

Ultimately, the particle physics of what goes on at those levels, down to quarks, is beyond the scope of this treatise. That would take years more to study and understand.

Yet, it is enough from the above calculations using just electrical-charge-force and magnetism-force that magnetism is stronger at the nucleus distance level; strong enough to hold a nucleus together at the distance of proton-neutron-proton separation, and yet charge-force could be much bigger and creates observed nuclear explosion without that separation if the proton come near to touching.

'Mass' Gets Replaced by a Surface Integral Over the Electron Shell Radius

This revision of mass into two components allows more calculations. It creates alternatives to 'having mass change' which has haunted the scientific community for years.

Infinite mass. A Stanford SLAC veteran scientist once said to me, "well, we all ignore that one." At the time, we were discussing a more interesting part of AVSC Atomic Model that particles getting accelerated near the speed of light, yet the magnetic field remains pointing in a different direction – perpendicular to the movement.

(106)

Particles, at SLAC, accelerate to near the speed of light within about four (4) feet within that massive facility, but then is takes a building of magnetics to direct that into a particular target location. The energy to move a particle is enormous. All that magnetics to move one particle. Mass must be infinite!

Most interestingly, the magnetic field of the near-speed-of-light particles are perpendicular to the direction of movement. The answer is part of the new theorem of what is the nature of nucleomagnetics.

Of course, what you have is a presentation of naked nucleomagnetics, the particle has gotten stripped from its atom so the energy is 1/distance-cubed.

The $\frac{1}{(x)^3}$ for $x = (R_{ES})$ reduction is gone. That is an increase of 10^{35} or 100,000,000,000,000,000,000,000,000,000,000,000 times increase. You can understand why anyone would think that is 'infinite mass', yet really that is naked charge or naked nucleomagnetics (which one is critical depends on the distance from other particles). It is not mass at all, but naked particle forces moving so quickly that the force to move it can barely keep up. Hence, the building worth of magnetics.

A similar naked force is the EMP (electromagnetic pulse) of a nuclear explosion. The short period of not surrounding electron shell can destroy every electronic system for miles.

With two components in AVSC, then we have the first part: the combination of protons and neutrons in the nucleus.

Number of Nucleus Particles is Non-Changing

Back to the calculations, with two factors create new opportunities and methods. It also has challenges.

The great leap of these revised definition of mass using the number of nucleus, the sum of the protons and the neutrons. That is concrete; anyone can count the number of items. We know the Atomic Count and Atomic Weight of each element. Carbon is 6 Protons, Atomic Count, and 12 Atomic Weight. Well, the number is 12.01 because 1% of these Carbon atoms have six (6) protons and seven (7) electrons for an isotope of 13 Atomic Weight.

Observed Mass is Directly Based upon the Sum of Protons and Neutrons

The second part is more complex. The radius of the electron shell something that changes. There are tables showing all of them.

One challenge is all those tables were presented based upon the Elements, the number of protons only. An amazing thing will happen when you look at this table by the isotopes. And then understand the calculation over that space.

Another challenge is that electrons sit at different distances, different angles, and thereby there are different distances. However, there is a great mathematician, Gauss. Gauss determined for any situation with the energy, the total force is determined by the special location of the last unit, the last electron. The organization of the inner bundles of energy fit as best they can; that is the electron subshells.

In Every Atom, Electrons Settle into positions Both Outside and Inside the Bohr Radius

Here is a table of the electron energies – which is a great substitute for the electron distance from the nucleus.

(107)

Element	Covalent Radius	Notes
01-H Hydrogen (1.001)	25 pm	
08-O Oxygen (15.999)	60 pm	
09-Fl Florine (19.998)	50 pm	
11-Na Sodium (22.989)		
15-SP Phosphorus (x.446)	100 pm	
16-S Sulfur (x.446)	100 pm	
17-Cl Chlorine (35.446)	100 pm	
26-Fe Iron (55.845)		
35-Br Bromine (79.901)	115 pm	
53-I Iodine (126.904)	140 pm	
54-Xe Xenon	216 pm	
84-Po Polonium (209)	190 pm	
88-Ra Radium (217)	285 pm	

That is the weirdest thing. As the nucleus gets 209 bigger, but the radius only grows 10x. In fact, from the full Shell-3, the Atomic

Number grows 5x, the Atomic Weight grows 7x, and the Covalent bond Radius grows 7x. Hmmm! Covalent bond distance grows at the same ratio as the Atomic Weight (the sum of the number of protons and neutrons). That seems very similar to our experience. The total particle count drives the Nucleus Size.

The Total Number of Nucleus Particles Directly Correlates with the Atomic Radius.

If you focus only on the Halogen Group, results show that radius is highly related to the Atomic Weight (the sum of protons plus neutrons):

(108)

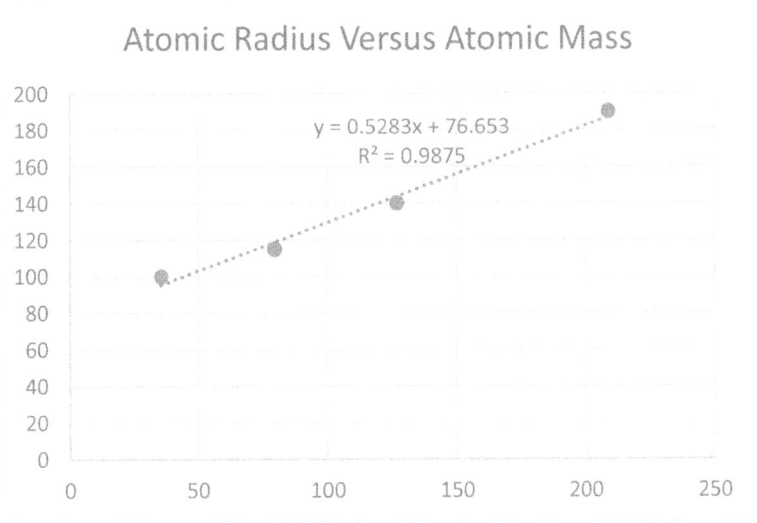

The average radius is consistent relative to the nucleus particle count. That means that we can Average Radius (R_{ES}), the Radius divided by the Atomic Weight. That average, when integrated of millions of particles means that R_{ES} does not change generally. It certainly does not change over an integral of millions of molecules of different types over enough time for those all to rotate, by their own heat, a few million times.

Configurations of Nucleus Chains/Rings

Basic Chain

The magnetics cause the top ring to flow P > N > P and so on. There is a magnetic pole (blue) and magnetic fields (red) around

A 001-H Hydrogen is easy because it is just one proton, so there are no other proton (+) particles to cause repelling. However, 001-H Hydrogen also has two isotypes which include neutrons.

(109)

001-H Hydrogen Trillium (3 Atomic Weight Isotope)

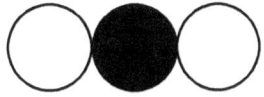

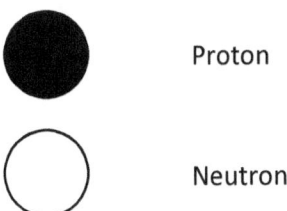

Proton

Neutron

This structure is stable. It meets the primary rule which is:

At least one neutrons must separate every pair of protons (+).

A nucleus could continue with this logic and create these long chains.

(110)

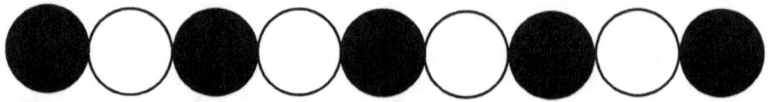

This structure has a long chain, and that makes the structure have a magnetic orientation.

(111)

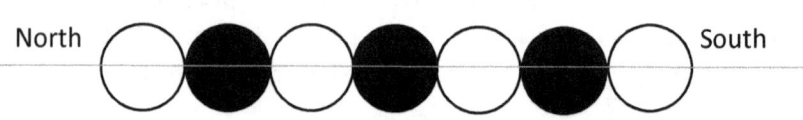

Remember that balancing act of 1) charge and 2) magnetics logic throughout this proof.

This would be great except that a chain is not rigid. A chain can bend around, and maybe those two ends meet again – and BANG a proton-proton nuclear-reaction repulsion.

(112)

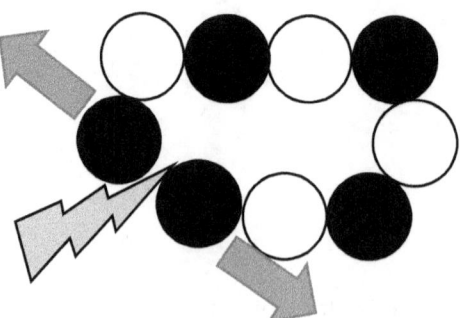

If a nucleus has a long chain and the two ends create proton-proton nuclear-reaction, which makes the chain not the most stable.

Magnetism-force follows the rule about attraction and repulsion similar to the charge-force except that the attribute is the direction of the magnet – north versus south:

-       Like Magnetic Poles Repel

    - Positive (+) to positive (+) repels
    - Negative (-) to negative (-) repels

-       Opposite Magnetic Poles Attract

    - Positive (+) to negative (-) attracts
    - Negative (-) to positive (+) attracts

So, if the two ends are a proton and a neutron, then 1) there is no proton (+)-proton (+) charge repulsion; and 2) the neutron adds magnetic attractions; which is 3) by the chain the attractive north-south orientation. Therefore, a chain if ending one end with a proton and the other end with an electron has a high tendency to flow around and become a ring.

(113)

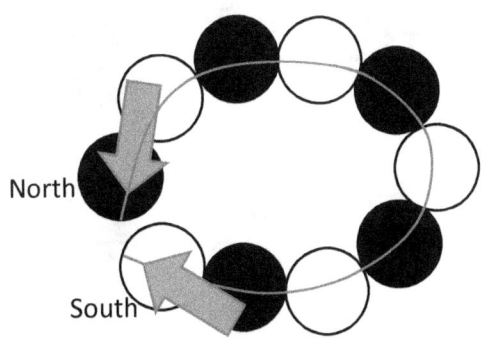

115

A chain must have one end with a north pole, and the other with a south pole. That means that, unless both end with protons, the two ends want to come together. Even if both ends are neutrons, the chain will still join.

(114)

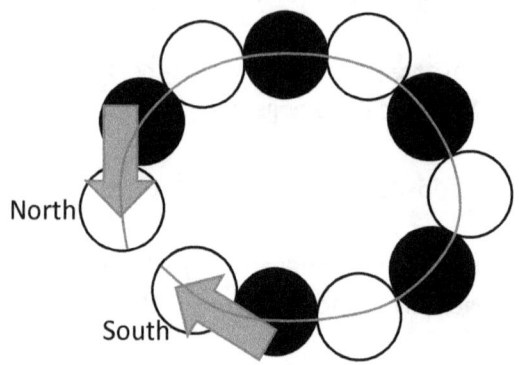

So, most often, the nucleus chain will still join into a ring. That structure is stable, and no endpoints can 'flap' and create the nuclear-explosion repulsion.

(115)

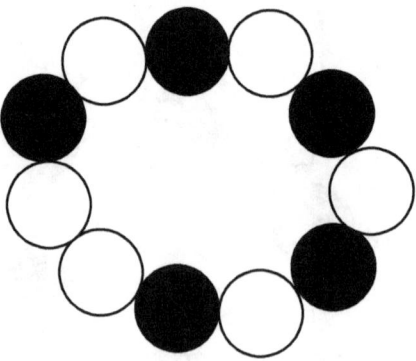

Finally, there can be combinations of rings and chains. The building block do not work neatly or perfectly.

(116)

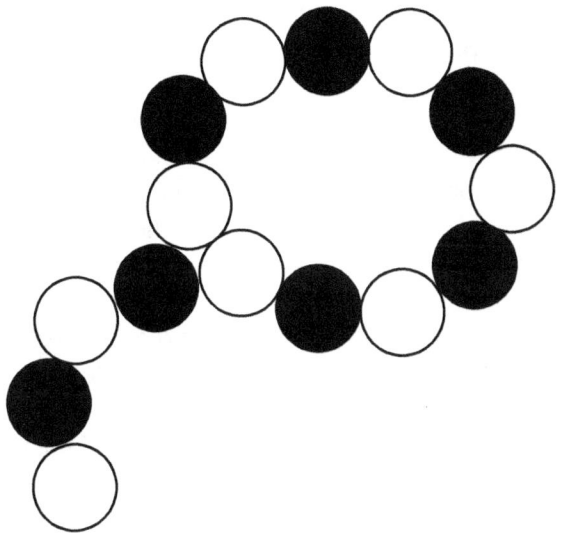

These combinations can also go in three dimensions (3D) which is more difficult to show on a 2D page.

These complex combinations become more important as you get to elements with lots of particles, especially those that are radioactive. However, the next chapters will focus on the stable lower end of the periodic chart with less particles. That will build the concepts, their attributes, and their impact on chemical structures, bonding and chemical reactions.

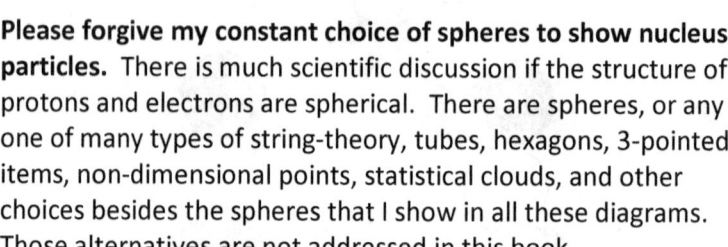

Please forgive my constant choice of spheres to show nucleus particles. There is much scientific discussion if the structure of protons and electrons are spherical. There are spheres, or any one of many types of string-theory, tubes, hexagons, 3-pointed items, non-dimensional points, statistical clouds, and other choices besides the spheres that I show in all these diagrams. Those alternatives are not addressed in this book.

I am using spherical particles simply because the pictures with spheres work to make ring angles easier. Please indulge me with that visual concept. I am not claiming that all nucleus particle are spheres; I just like them for the picture of the ring-concept.

Much research describes small-number-of-particles atoms have a nucleus size at $0.8 * 10^{-15}$ m, yet larger-number-of-particle atoms have a nucleus size at $1.5 * 10^{-15}$ m suggests that the distance between nucleus particles, and thereby the radius of individual particles, or subparticle clouds, actually compresses. Given that 1.5/0.8 ~= 2 or twice as big, you cannot have the 200 particles of a large atom fit into a volume that is only $2^3 = 8$ times bigger if the particle remains the same spacing or size. That is, the particles get into a structure which makes them fit together tighter.

My current thinking is that the volume of larger atoms for 200 particles shows the basic structure. So, at $(1.5 * 10^{-15})^3$ m^3 / 200 particles =

$$V = \frac{(1.5*10^{-15})^3}{200} m^3 = \frac{3.4*10^{-45}}{200} m^3 = 1.7 * 10^{-47} m^3.$$

Then taking the cube-root of that is the core radius of $0.3 * 10^{-15}$ m^3. This corresponds with other observations of the neutrons which states it has a positive-charged core of the same $0.3 * 10^{-15}$ m.

None of that actually resolves sphere versus cone versus string versus subparticle cloud versus something else. Therefore, I stay with sphere until something is proven.

Secondary Understanding: There can be two or more neutrons in a row.

While the perfect nucleus would be these proton and neutron in perfect order, that is not reality. Depending on the elements, you are almost as likely to have either chains or rings with places with more than one neutron together – still separating any repulsions from protons (+) touching or near-touching protons (+).

(117)

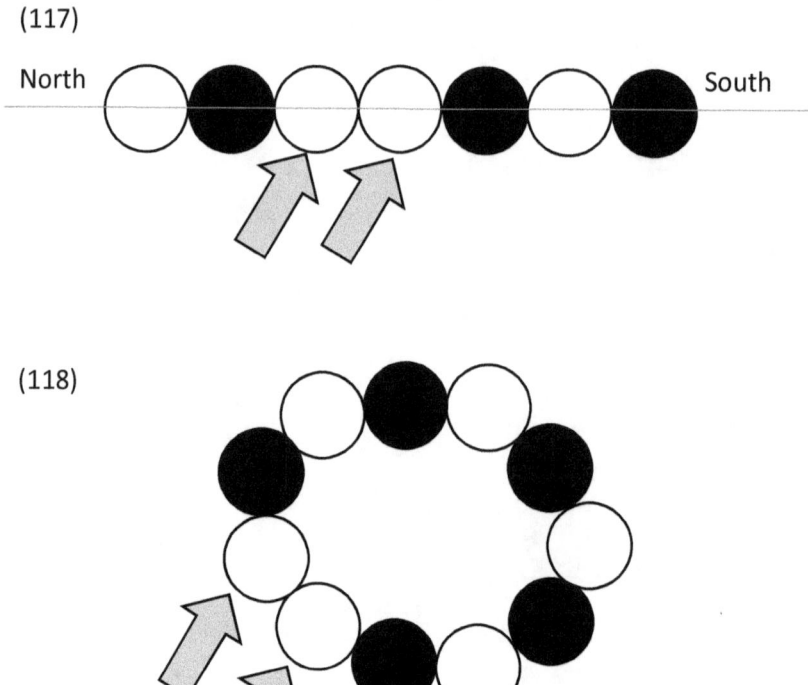

(118)

Remember this fact because it explains various forms of the same element called isotopes. That is where an element can exist with the same number of protons, but with a different number of neutrons. We will explore these isotopes more in a later chapter.

Magnetics of Entire Nucleus Structure Are Important and Different than Particle Magnetics

The basic concept of a ring of magnetic particles creates a 2nd overall magnetic direction. There is a magnetic pole at 90 degrees to the ring.

The magnetic particles many orient with each south connecting to the north of the next particle. However, this creates a secondary magnetic field that has it poles through the middle of the ring. The particles have a magnetic orientation that flows like the blue arrow. However, magnets in a ring create another north-south magnetic field at perpendicular (the red line).

(119)

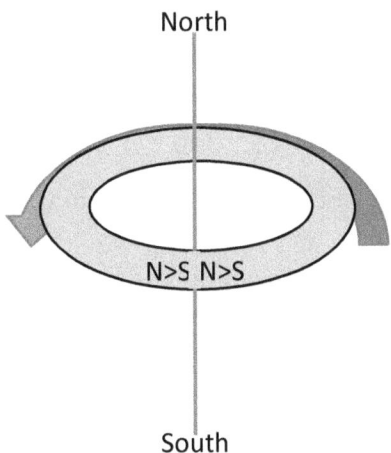

And the fat part of the filed around the ring-circle, sort of like a 'bagel' or maybe a series of these loops of magnetic field surrounding the ring. That field being strongest around the ring.

(120)

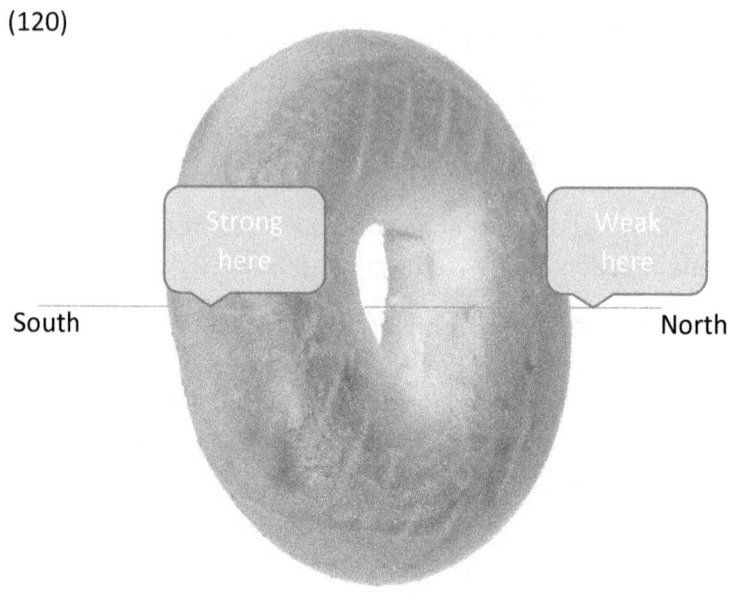

South North

(121)

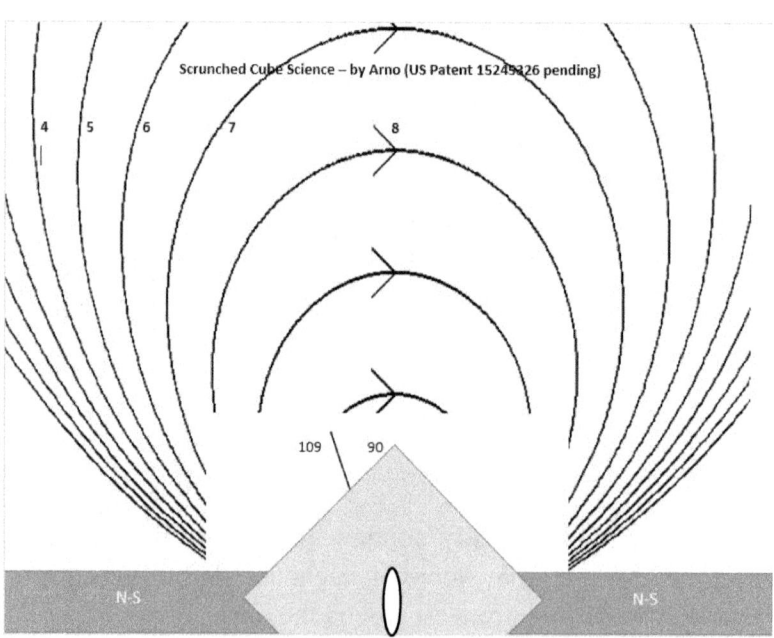

122

However, a strange thing happens when if the nucleus builds into two rings. The magnetics of the 'lower' ring reverses. To work, the upper must be N>S, and the lower S>N with the connection as little loops of magnetic energy.

(122)

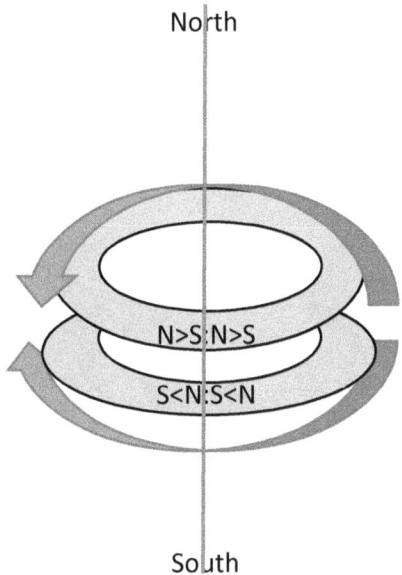

The magnetic field between rows becomes a series of loops.

(123)

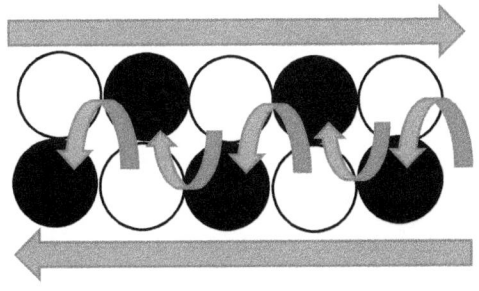

Single-ring:

The magnetics cause the top ring to flow P > N > P and so on. There is a magnetic pole (blue) and magnetic fields (red) around until a ring completes.

(124)

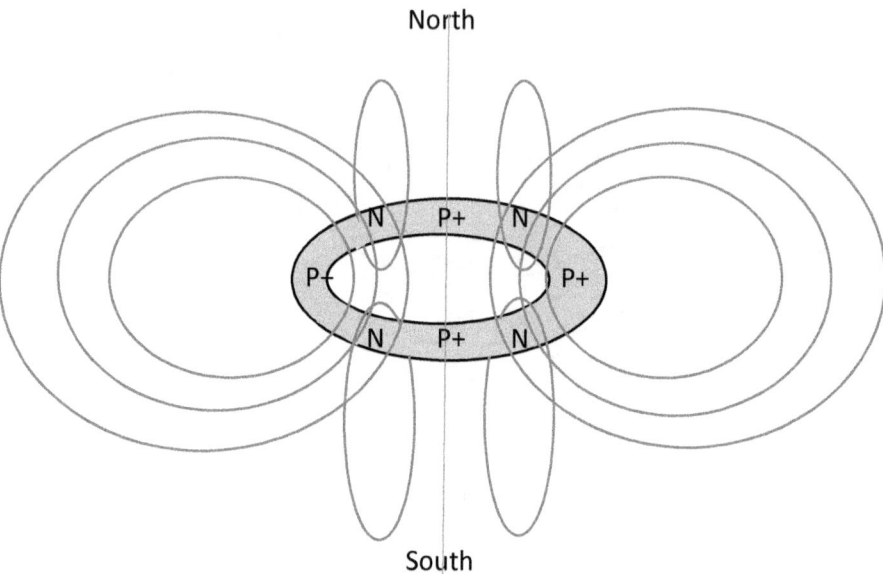

Double-ring:

The magnetics cause the top ring to flow P > N > P say in N > S
(125)

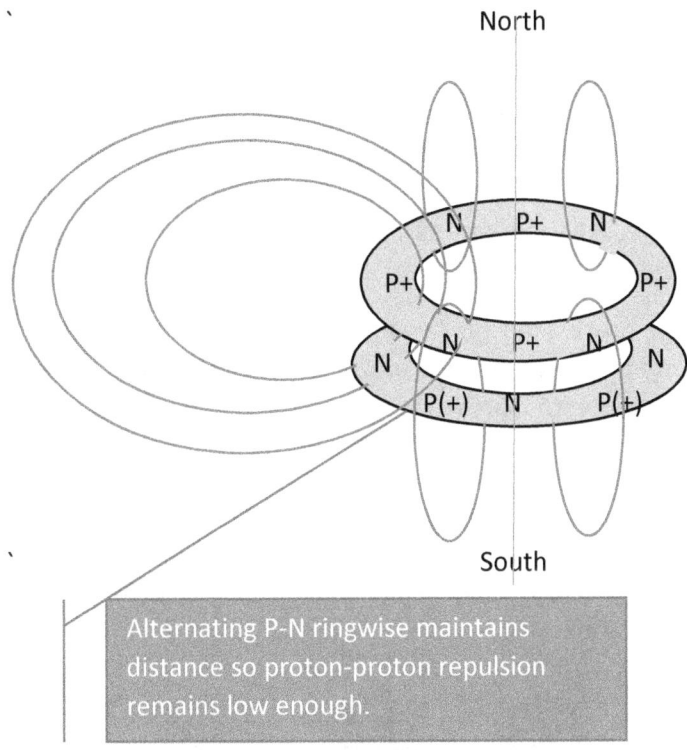

Alternating P-N ringwise maintains distance so proton-proton repulsion remains low enough.

But because of the connection from layer 1 to layer 2, the polarity changes for the 2nd layer

The magnetics cause the 2nd ring to flow N > P > N say in S > N. That is the flow of N-S is the opposite.

Building Higher Elements Needs Neutron First to Separate for Added Proton

The building of chains would seem easy. You add a neutron, then a proton fits in between. However, this of how long and wiggly this structure becomes. In fact, too much wiggly, and the chain will bend, and if a proton nears a proton, the structure explodes. Yes, a proton-to-proton touch or near-touch is the basis of nuclear explosions. Anytime that protons get too close the repulsive power becomes huge, likely tearing apart one section of the atom's nucleus from the rest.

The process must follow a particular path of a) add neutron first, then, add proton to keep the proton-proton near-touch decay of the entire structure from occurring.

The Stages

A chain (segment of nucleus particles) is naturally proton-neutron-proton and so on. (126)

A neutron can attach to that segment. A neutron protrudes.

(127)

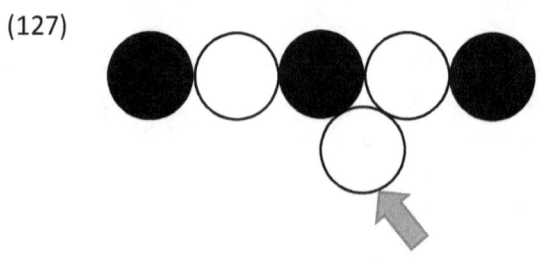

However, then the particles realign so a neutron-neutron block occurs.

(128)

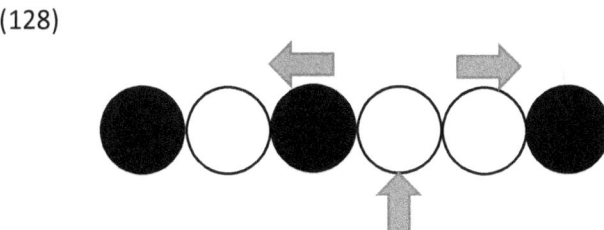

Only after that realignment, a proton can attach. A proton protrudes.

(129)

Of course, those can then re-orient back into a longer chain (and new element).

(130)

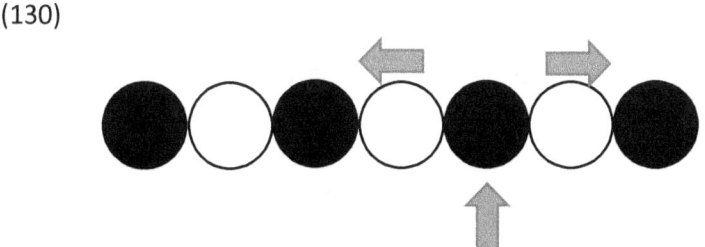

In the physical world, you will find a nucleus in each configuration. Effectively, a nucleus can be at any of the stages for any segment. That is why larger molecules are not perfect 2:1 ratio. It can work with a small structure, but as the structures get bigger some of these neutrons gets stranded.

(131)

	Atomic Number	Ratio
002-He Helium	4.0026	2.0013000
006-C Carbon	12	2.0000000
012-Mg Magnesium	24.3	2.0250000
092-U Uranium	238.0	2.5869565

In Multiple-Ring Structures, Extra Neutrons Keep the Protons Separated

In fact, to get to the multi-layer rings, there needs many extra neutron spacers so protons do not touch. Once an atom gets into multiple rings, the process is more complex. The 2nd layer cannot have protons in the same positions as the initial row.

(132)

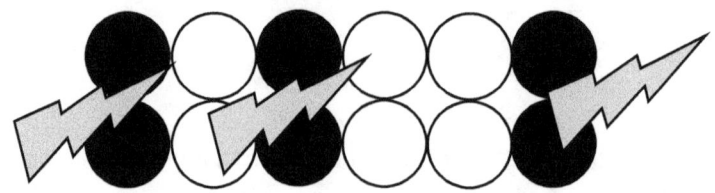

That would create a nuclear explosion. However, if those align perfectly, then a 2-layer double ring works.

(133)

Side View

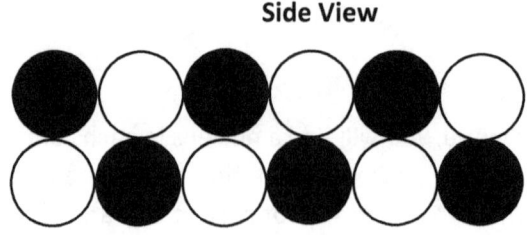

However, in nature, the more likely building includes some spacers. That is, a neutron sits in a spot every so often because that can happen as easily as a proton (+).

(134)

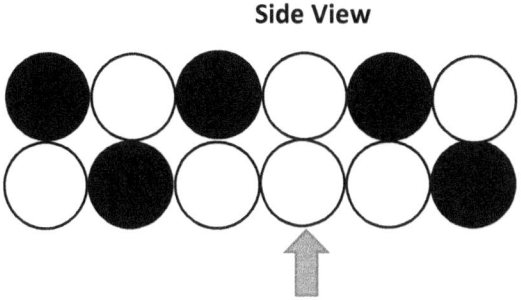

Structure Prefers Same Number of Particles on Every Layered Ring

If two layers have the different number of element, then magnetics like to drive the structure into the slots. However, that makes every particle from one ring touch two particles in the next layer. That makes it almost impossible for a proton not to get near another proton. Of course, that, BANG, would create nuclear decay.

(135)

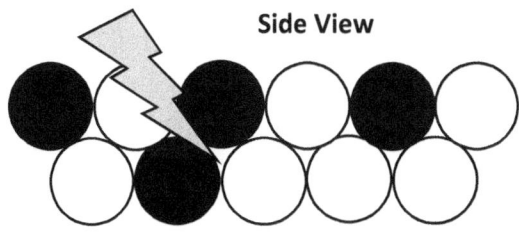

If there is an odd number of particles in each ring, then definitely an extra neutron fits into the structure to make the structure have the same number of elements on each ring-layer.

009-N Nitrogen has an isoform-isotope which has five protons (+) on one layer and four (4) protons (+) on the other layer of a 2-layer ring structure. But that does not work. The firth positions shifted alternating would become proton at each end where the rings connect. Instead, one common isotope one extra neutron makes the structure stable. This is a P-9, N-10 isotope.

(136)

Side View

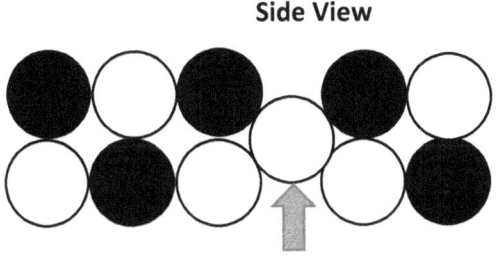

(137)

Top View

The spacer is important because otherwise, one of the layers might have one end with a proton touch the other end also with a proton.

Evidence:

Look the preferred isotopes of 026-Fe Iron. In nature, it seems it is never built perfectly 26 protons and 26 neutrons = 52 particles. Instead you get frequency of the isotopes of:

026-Fe Iron Isotopes	Frequency
53	0.05%
54	5.90%
55	0.05%
56	91.72%
57	1.20%
58	0.28
Total	100.00%

The most preferred for an even number of protons is also even. Even more preference in multi-rings even elements is that isotope has a multiple of four in total nucleus particles. My interpretation is that the best structure is two layers of twenty-eight (2 x 28 = 56) or four layers (4 x 14 = 56). The other favored isotopes are probably (2 x 27 = 54). But the true odd ones are not often chosen.

Something different happens when the number of protons start out as odd. Remember that the extra proton can sit at the end of a protrusion or to make the rows match a single neutron can separate and lock two odd-number rows in ring.

(138)

Side View

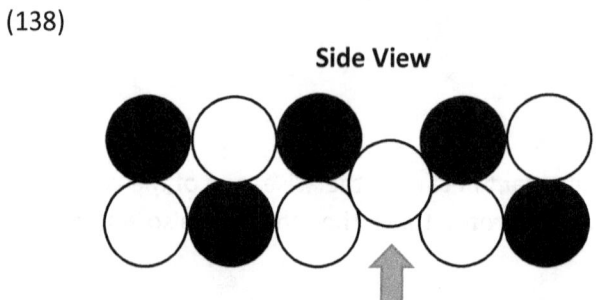

There must be a full substitution to get the rings the same. That means an extra neutron holding two rows, so odd. Or extra neutrons with that proton in a protrusion.

027-Co Cobalt Isotopes	Frequency
57	0.00%
58	0.00%
59	100.00%
60	0.00%
Total	100.00%

Charge Repulsion Causes Need for Extra Neutrons in Various Isotopes – Building Larger Elements

This separation of the neutrons means that only one path works to build a larger nucleus. You must add the neutron first. That creates two neutrons together. Then a proton can attach between those two neutrons safely. Eventually, a path to a higher element (more neutrons) occurs, but only in that order. Otherwise, protons (+) touch and nuclear decay of the structure.

A chain (segment of nucleus particles) is naturally proton-neutron-proton and so on. (139)

A neutron can attach to that segment. (140)

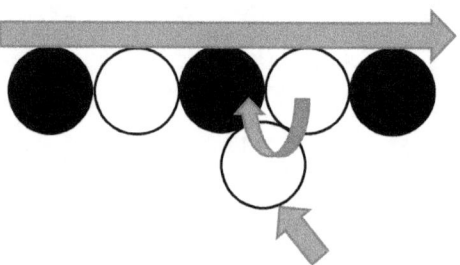

However, that structure creates an extra magnetic loop besides the straight north-south direction of the magnetic chain. This loop is a force holding that new neutron as a protrusion.

However, if a force works to break that magnetic loop, then the particles realign so a neutron-neutron block occurs.

(141)

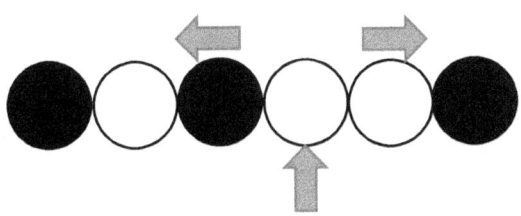

Only after that realignment, a proton can attach. Of course, the magnetic field for this intermediate stage also has a strange loop. Remember that must apply force in order of strength:

1) Charge is strongest. In this case, proton-proton repulsion

2) Magnetic is next. We must break the extra magnetic look before there is room for the next neutron.

(142)

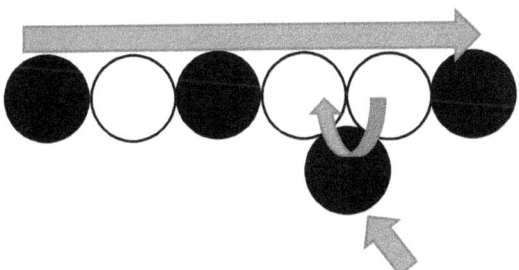

Of course, those can then re-orient back into a longer chain (and new element).

(143)

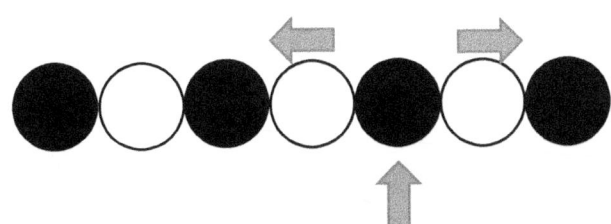

We see this in the radioactive element 088-Ra Radium. The structure is too big that a proton will eventually touch or near-touch each other and the element will change:

The number of protons is conserved:

(144)

Element Before	Elements After
088-Ra Radium	086-Rn Radon
	002-He Helium
Total # of Protons = 88	Total # of Protons = 88

The total number of nucleus particles is also conserved:

(145)

Element Before	Elements After
088-Ra Radium (isotope 226)	086-Rn Radon (222 isotope)
	002-He Helium (4 isotope)
Total # of Nuclear particles = 226	Total # of Nuclear particles = 226

Why use Magnetics versus Charge (kQQ)?

The secret of the gravity calculation is that is based upon the point where electrostatic charge (Coulomb's Law kQQ) equals the nucleomagnetics force M(Z_1+N_1). Everyone knows that gravity decreases like the electrostatic charge 1/distance-squared.

At the balancing point (RES), those two balance, so I could just as easily state the formula as:

(146)

$$E = c^2 * \frac{kQQ}{2*\frac{4}{3}\pi(R_{ES})^3} \quad \text{OR} \quad E = c^2 * \int \frac{M(Z_1+N_1)}{2*\frac{4}{3}\pi(R_{ES})^3}$$

However, the radius of the electron shell (R_{ES}) is constant when using the Nucleomagnetics force. The integral of the Radius varies if you just use the electrostatic charge.

Notice that the secret of AVSC gravity transformation is that the charge force attraction electron-proton equals the magnetic repulsion.

The magnetics adds a volume of extra repulsion. Volume = 3 dimensions. Yet, it decreases a 1/distance cubed (1/d^3). Electrostatic charge adds repulsion at 1/distance-squared (1/d^2) which is not a volume. So, I cannot get the average/integral as linear, like with magnetics. As such, the measurements of 'mass' are consistent by the magnetics particles, yet charge does not work well.

Why the extra Neutron in some elements?

In the most common isotopes of these elements, some configuration like an extra neutron while others do not tend to contain the extra neutron: (147)

Element	Atomic Number	Protons	# of Neutrons
003-Li Lithium	6.941	3	4
004-Be Beryllium	9.0122	4	5
005-B Boron	10.811	5	6
006-C Carbon	12.0107	6	6

Many times, the structure works with a standard ring, P-N-P-N- until back to the beginning. However, some structures work better with an extra neutron to create stability. 006-Carbon is a perfect 2:1 with twelve (12) nucleus particles for six (6) protons. For 006-C Carbon, we find N = P, so # = 2P or 2.000. Yet, the other elements are N = P+1.

Take for instance, Beryllium, with four (4) protons and the most common isotope with five (5) neutrons. This should only need four (4) neutrons to separate the four (4) protons into a ring. However, what happens more often is that an extra neutron arrive and one end of the ring become fixed. A ring eight around is extremely flexible.
However, adding another neutron, and you get a box-and-diamond combination that is much less flexible.

(148)

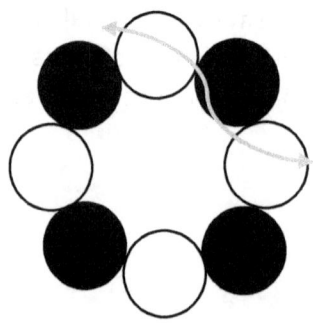

(149)

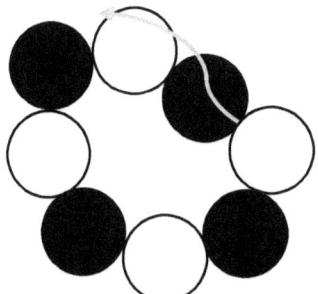

The above has lots of flexibility, maybe even too much so the protons would get too near each other. The below is very rigid with an extra neutron added. The added neutron makes the ring not able to twist. The magnetic double connection via the extra neutron makes this structure rigid.

(150)

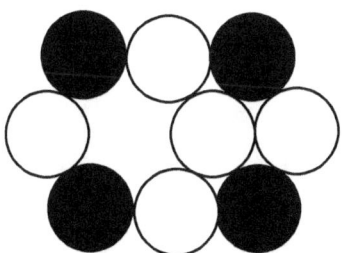

Therefore, you get a tendency for the 004-Beryllium to have five (5) neutrons.

(151)

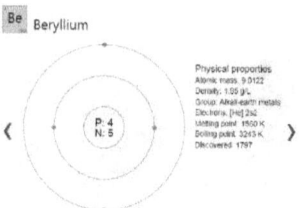

This give 004-Berylium a square proton configuration in the nucleus. That has a gentle effect on the electrons, and thereby on bonding.

Why are certain large atoms radioactive? And why consistent decay?

Because at large numbers and size, the combination with neutrons separating protons eventually have too many chains attached, and at a consistent rate, the chains touch another proton which makes them separate (decay) into two particles. The perfect rings I describe of small number of particle atoms are not common for the largest, radioactive atoms.

The large atoms are mixes of multi-layer rings with chains attached.

(152)

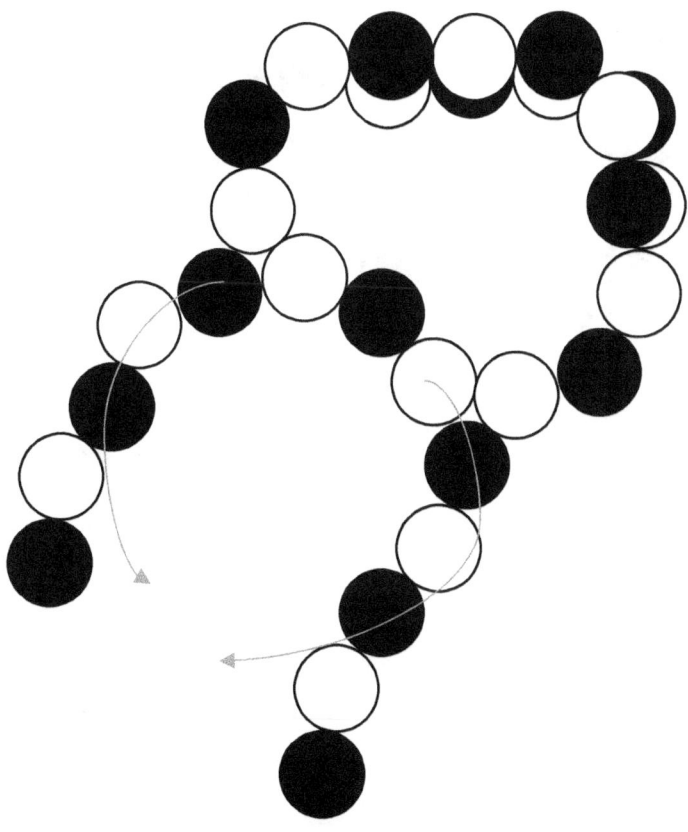

And, it two of chains get long enough and flop into each other, and that is radioactive decay.

(153)

The ratio is consistent, and as they meet and decay, the number goes down, exactly as per the radioactive decays charts.

Visualization of Nucleus Chained/Ring Configurations for Various Elements

Here are pictures (using small magnetic balls to visually simulate with red as the proton and green as the neutron) each element and its isoform (ring-configuration): (154)

Element	Neutrons	Structure Form	Picture
H = Hydrogen	0	No ring – N/S	
001-H = Deuterium	1	No ring – N/S	
002-He Helium			

Element	Neutrons	Structure Form	Picture
003-Li Lithium	3	Single ring	
004-Be = Beryllium	4	Single ring	
004-Be = Beryllium	5	Single ring	
004-Be = Beryllium	5	Locked ring	

Element	Neutrons	Structure Form	Picture
005-B = Boron	6	Single ring	
006-C = Carbon (coal)	6	Single ring	
006-C = Carbon (graphite)		Linked Rings	
006-C = Carbon (diamond)		2-layers of rings	

Element	Neutrons	Structure Form	Picture
007-N Nitrogen	8	2-layers of ring 1-proton sticking out	
008-O Oxygen	8	Single ring locked	
011-Na Sodium		2-layers of rings	
012-Mg Magnesium		2-layers of rings	

Until you have tried to make magnetic chains fit in a certain way, and kept the protons apart, you cannot understand the ways in which the loops of magnets can only build in certain ways.

Try it!

For example, you can try to create two lines:

(155)

Side View

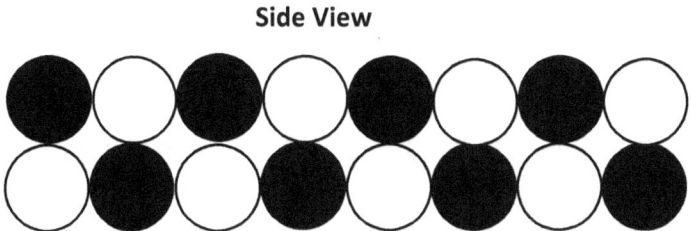

However, the magnetic forces cannot rotate at the ends. The magnet will magically move back to loops at the end.

(156)

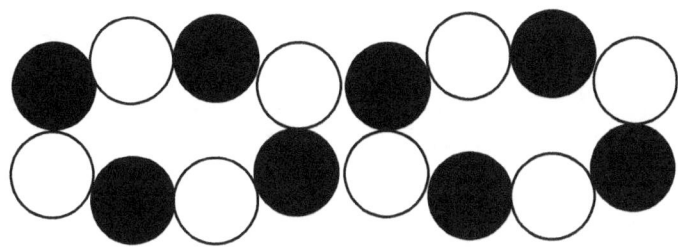

For the advanced explorer, you can have calculated why those ends cannot make the 90, then 90 degree turns.

For the people that learn by experience, try it!

Multi-Layer Rings Based upon Number of Particles (by Element)

As I showed, the nucleus can arrange in any way. However, from the magnetic moment calculation, you can guess that certain configuration is most common based upon the number of particles.

(156)

Element	Protons	Neutrons	Structure Form
001-H - Hydrogen	1	0	1 particle
001-H - Deuterium	1	1	chain – N/S
001-H - Trillium	1	2	chain – N/S
002-He - Helium	2	2	Single ring
Li - Lithium	3	3	Single ring
Be - Beryllium	4	4	Single ring
B - Boron	5	5	Single ring
C - Carbon (coal)	6	6	Single ring
C - Carbon (graphite)	6	6	2 flat rings - Infinity-8 loop
C = Carbon (diamond)	6	6	Double ring
C13 - Carbon+1Neutron (coal)	6	7	2 flat rings - Infinity-8 loop
C14 - Carbon+2Neutrons (coal)	6	8	2 flat rings - Infinity-8 loop with extras locking in center
N - Nitrogen	7	8	Double ring
O - Oxygen	8	8	Double ring
F - Florine	9	10	Double ring
Ne - Neon	10	10	Double ring
Na - Sodium	11	11	Double ring

Mg - Magnesium	12	12	Double ring
Al - Aluminum	13	13	Double ring
Si - Silicon	14	14	Double ring
Ph - Phosphorus	15	15	Double ring
S - Sulphur	16	16	Triple ring
Cl - Chlorine	17	17	Triple ring
Ar - Argon	18	18	Triple ring
K - Potassium	19	19	Triple ring
Ca - Calcium	20	20	Triple ring
	21	21	
	22	22	
	23	23	
	24	24	
	25	25	
Fe – Iron	26	26	
Co – Cobalt	27	27	
Ni – Nickel	28	28	
Cu - Copper	29		

Isotopes and Isoforms

For a century, there have been a concept of isotopes. That is a class of nucleus with the same number of protons, but grouped (as isotopes) based upon the number of neutrons.

However, the proposed structure allows nucleus to vary in other ways. You can have nuclei with the same both protons and neutrons, as such the same isotope, but the physical structure is different.

(157)

002-He Helium (4 Isotope – chain isoform)

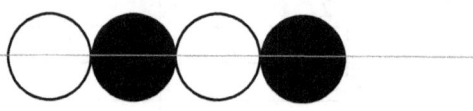

(158)

002-He Helium (4 Isotope – 1-ring isoform)

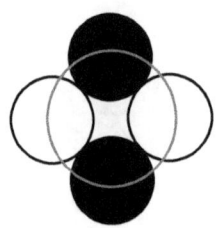

Both of these have 2 protons and 2 neutrons = 4 nucleus particles.

However, the magnetics of the two are very different. The 1-ring has lower strength, and a magnetic pole orientation at 90 degrees to the ring. However, the chain isoform has stronger magnetics and a magnetic orientation directly in line with the magnetics of the individual particles.

Further, the linear isoform remains very flexible. The 1-ring structure is very rigid.

The same happens with isotopes and isoforms of 006-C Carbon (C-14 Isotope – 1-ring isoform extra neutrons outside) is flexible.

(159)

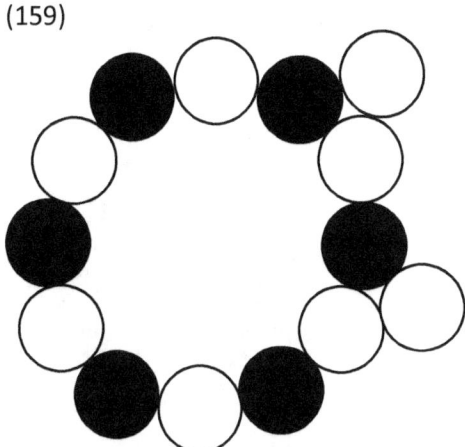

But the same isotope is more rigid if the extra neutrons bind inside.

(160)

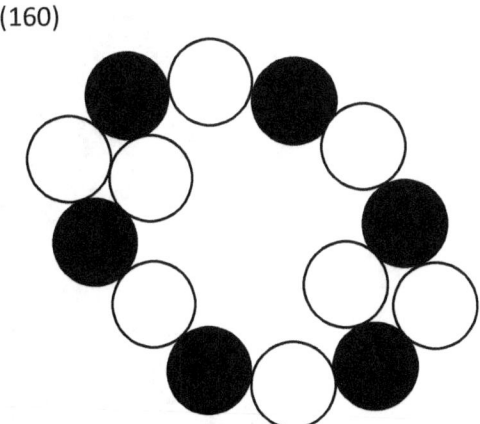

Proposal: Transparency and the Nucleus Structure – 006-C Carbon as Coal versus Diamond

In case you missed, I showed three isoforms of 006-C Carbon.

(161)

006-C = Carbon (coal)			
006-C = Carbon (graphite)			
006-C = Carbon (diamond)			

If the structure of the nucleus is flexible, then it can absorb energy easily. Lots of frequencies can move the particles, and thereby do not move on.

(162)

However, a 2-ring structure is more rigid.

(163)

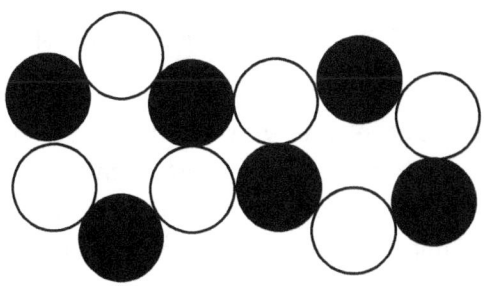

However, a 2-layers-of-rings structure is extremely rigid. Nothing can move.

(164)

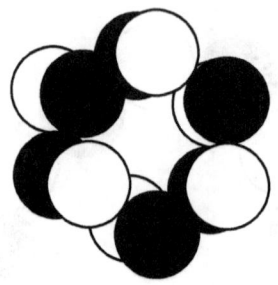

(165)

Isoform	Flexibility	Range of frequencies absorbed	Transparency
1-ring	Lots	Up to	None (black)
Flat 2-ring	Limited	Only small range	Silvery
2-layers of rings	None	Almost none	Transparent

The color of these absorption or transparency is based upon the diameter of a nucleus.

There is another source of color absorption from the structure of the crystalized bonding. If the electron bonds are oriented in a crystal, that covers absorption or translucency of frequencies in the range of that distance. That limits the frequencies by this locking.

My postulate is that you have both types going on in clear crystals.

However, the clearness, clarity, to see through you need all frequencies limited in absorption, at both nucleus, electron shell, and bonding.

Fundamental Question: What is a Neutron?

The current research on the neutron focuses on observations that find the field around a neutron is not actually neutral. It is positive on the inside, and negative on the outside – which nets to zero.

The postulate is that a neutron is a proton with an electron attached by a neutrino.

Experimental evidence shows that when hit, a neutron can degrade into those three components: a proton, an electron, and a neutrino.

So, let's start at the beginning

How do a proton and electron bond into a neutron at very low temperatures?

There are two fundamental forces between a proton and an electron:

Charge attracts the two (charge-force opposites attract)

Magnetism repels the two (electrons flee all magnetics, north or south)

Glossary

Atom – The balanced combination of a nucleus consisting of P protons and N neutrons, with a shell of P electrons.

Bohr radius – The ratio at which the magnetic field and the charge-force field balance based upon a single-proton 001-H Hydrogen atom.

$$5.2917721092171717 \times 10-11m$$

Coulomb – the measurement of the charge-force of proton or electron particles.

Coulomb's Law

$$F = k_e \frac{q_1 q_2}{d^2}$$

Electron – an atomic particle with a negative (-) charge with a repulsion to magnetic fields.

Isotope – a class of atoms with structure of a nucleus which have the same number of protons, but not the same number of neutrons.

Isoform – a class of atoms with structure of a nucleus which have the same number of protons AND the same number of neutrons.

Magnetic field – the north-south oriented field associated with a proton or neutron. The grouping magnetic field may have different orientation, strength than the magnetic field of the individual particles.

Magnetic ring - A structure where magnets link north-to-south in a chain then connected at the distant end. As such, that field actually creates a further magnetic field perpendicular to the ring.

Neutrino – an atomic particle without charge and possibly without magnetism.

Neutron - an atomic particle without charge and with magnetism. Currently research indicates it may be a combination of subparticles.

Nucleus particles – particles that bind by magnetics in the nucleus. These are positive-charged (+) and no-charge neutrons.

Proton - an atomic particle with a positive (+) charge and with magnetism.

Four Arno Vigen Science Postulates:

For more than a century, the pendulum of physical sciences moved away from Newton and the concrete, physical world. The four Arno Vigen postulates below move the pendulum one step back towards center. A is A. Physical reasons are better describers of physical science, when those physical factors are discovered.

While other solutions get the correct answer in brilliant, amazing, creative formulas, the deep answer becomes simple and real. It becomes something that we can teach to every student, without LaGrange, Hamiltonians, and Gaussian differentials -- without, or better said resolving into the basics, the invented 7, 9, 11 or 18 dimensions of the latest version of string-theory.

The four postulates go back to the basics:

- Three physical dimensions (length, width, height or their spherical equivalents)

- Time

- Electrostatic and nucleomagnetics fields*

- Known particles – protons, electrons, and neutrons

These, 1) electric-charge-force and 2) nucleomagnetics are understood as intimately linked, but the full details of that linkage will wait until someone discovers the direct interconnection in physical dimensions.

Each of the postulates takes current, complex calculations, and gives them a clean path using only the above basics.

#1 - Every nucleus holds together via a chain/ring-magnet organized as proton-neutron-proton-and so on:

- Resolves what holds the nucleus together. This 'strong force' gets based upon when alternative particles, in a chain, connect along an oriented magnetic field, so that, at the nucleus distance, magnetism is stronger than charge-force if the positive (+) charged protons are separated enough by an intermediate neutron.
- Educational nucleus-plus-chemistry set patent 15256865 pending relating the magnetic field of the particles to the overall magnetic field of the chain or ring nucleus structure.

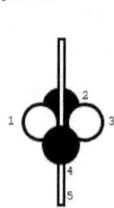

Figure 12

While Charge is stronger than magnetism, yet a nucleus stays together:

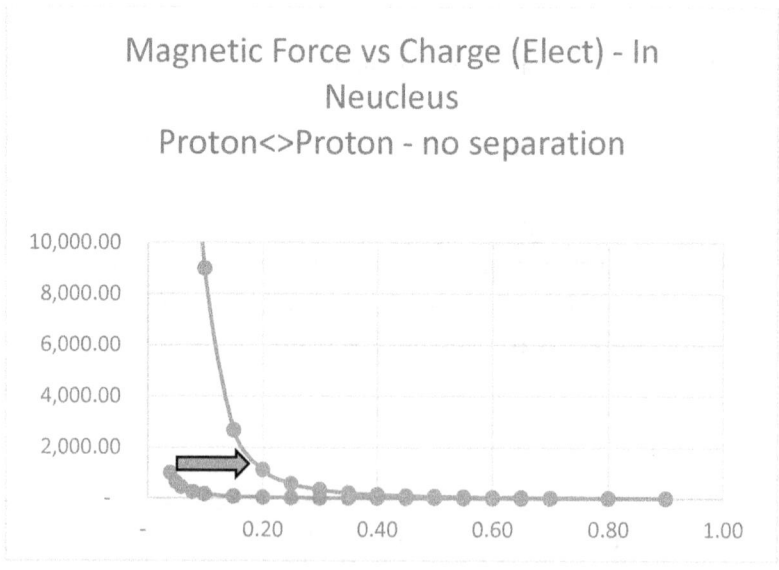

Because a magnetic field does not start decreasing until the physical chain is broken, there is proton-neutron-proton configurations that allow a nucleus to stay bonded together.

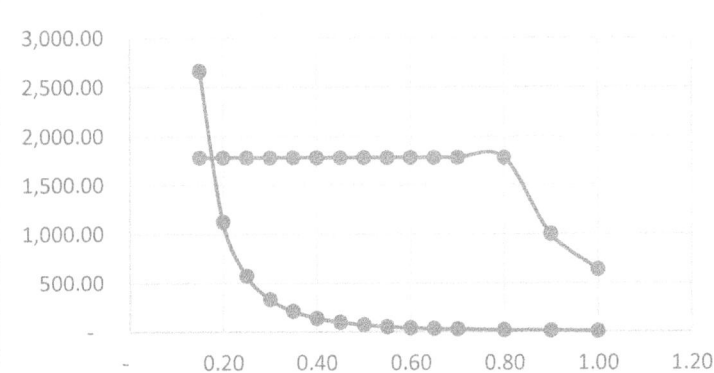

It simplifies to the below two graphs that explain the charge force versus the magnetic force. Both decrease, but for a point, the charge force is always stronger (chart 1), but in chains the charge is not a point, but a chain which means it does not decrease until the end of chain – keeping it strong enough to stay linked even if protons repel.

#2 Electrons repel magnets – *both poles*
- Resolves what force makes the electrons stay in a shell
- Resolves spin number in various subatomic particles
- Resolves color factor in various subatomic particles
- Explains basic electricity
- Explains EMP pulses

#3 Electrons shells build in geometric forms within the nucleus magnetic field with a) N-S doubling, changing by math, and not in the direct filling order of aufbau/Pauli

- Educational chemistry set patent 15245326 pending
- Resolves 006-C Carbon 109.5 angle versus 007-N Nitrogen 107.5 angle versus 008-O Oxygen 104.5 angle
- Resolves 027-Co Cobalt melting point
- Resolves 005-B Boron bonding angles at 120 degrees
- Resolves the 029-Cu Copper and other transition metal electromagnetic spectrum evidence why only 1 4s electron
- Replaces s/p/d/f with geometric m/e/c/t/v with e as intermediate in some elements. m2 = magnetic poles scrunched, e = equatorial, c6 = rest of cube with m2, t6 = tetrahedral blocks towards the magnetic poles, and such

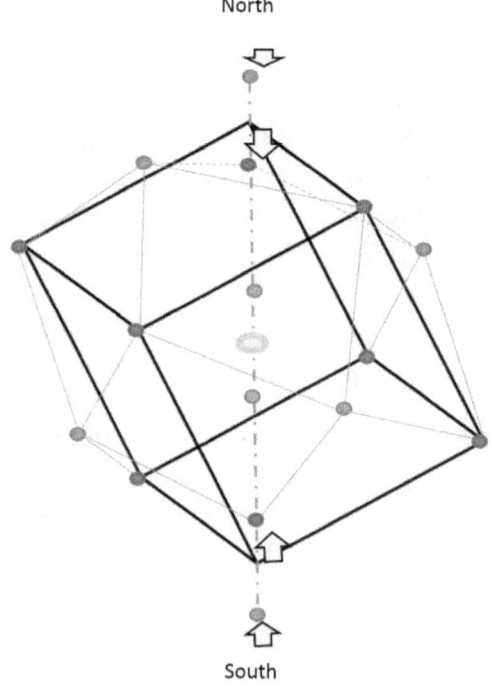

#4 Gravity is the nucleus (proton, neutron) magnetic field reflected via an almost* universal electron shell radius (R_{ES}).

- Permanently links gravity to the basic electromagnetism
- Resolves link of nucleus 'atomic mass' particles to 1/distance-squared observations of gravity
- Resolves mass loss in bonding by the bonds masking of R_{ES}.
- Replaces many of the time-space warping calculations with physical factors that change R_{ES}, a physical distance so the calculation is knowable with physical dimensions

$$Strgth = \frac{-M_A{}^2 n_1 n_2}{d^3(\frac{4}{3}\pi R_{ES}{}^3 + T)} = \frac{-(M_A n_1)(M_A n_2)}{d^3(\frac{4}{3}\pi R_{ES}{}^3 + T)}$$

$$mass = \sqrt{G} \; m = \frac{Mn_1}{2 * \frac{4}{3}\pi (R_{es})^3 + T}$$

*'Universal' will still have various Einstein, Lorentz adjustments as discussed, integrated, and potentially improved in other books now understood because based upon physical distance (R_{es}) which stretches, instead of space-time, as you get to a percentage of speed of light, and other factors.

What is the Energy of a Photon Applying Planck-Einstein?

Taking the original E=mc-squared for the Free Energy for a photon, the Planck-Einstein relationship calculate the energy of a photon as variations of:

$$E = hf = \frac{hc}{\lambda} = \hbar\omega$$

The most interesting in the application of these forms, as the wavelength, λ, gets very small, that would make the Energy near infinity. As λ > 0, then 1/λ > infinity. At the 'Planck distance', 10^{-26}m, the energy matches the maximum allowed by Einstein's Free Energy E=mc-squared.

Relativity will state that at that point it becomes a black hole. Of course, that does not have a physical model.

However, in AVSC, we think about that problem differently at two levels. First, at the physical model, AVSC works to explain the geometry:

- The energy of the photon is nucleomagnetics, and has physical orientation.

- That means a photon rotates – hence frequency and wavelength based upon an actual physical model. Because the force is toroidal, that rotation creates the known electromagnetic pulsation.

- As it rotates faster, it has more energy. Hence, hc/λ makes sense. When that photon is captured, it provides more

- That rotation also creates a motomagnetics field based upon the movement within its own physical dimensions.

- When it gets to limit, that rotation become field as powerful as nucleomagnetics, or maybe that is nucleomagnetics. As such, you cannot get the physical size of the field (a Planck) moving faster than the underlying field, the speed of light. This might be the physical model for how protons and electrons get built.

- So, when you rotate too much, it become charge without nucleomagnetics – much like a neutron become nucleomagnetics without charge.

- When the receiving has a harmonic frequency, it absorbs this rotating nucleomagnetics energy, and in turn has its own higher energy which in turn can push it to a new settling positions (absorbing or generating spectrum emissions).

This leads to further postulates on the grand scale:

- Dark matter, photons, and probably neutrinos all have nucleomagnetics, but not electrostatic force. These are like a neutron that is likely a combined electrons and electron with electrostatic charge balanced, but nucleomagnetics still strong.

- A photon is rotating nucleomagnetics phenomenon. That is what gives it the specific frequency (ω). A photon is only captured at the distant end by something with the same frequency at the correct time and location. Otherwise, the repulsive energy rotates the nucleomagnetics field that the particle is deflected during an out of synch rotation.

- If an approaching photon rotates at one rate, it can

What does it mean that the energy of a photon changes by the frequency?

When the photon arrives it is rotating at a frequency (ω). It has the same amount of nucleomagnetics total energy ($\hbar$).

Considering Newton, the total momentum of the system, a) when separate beforehand, and b) when combined afterward, must remain constant.

At low frequency, the energy would be added when they synch:

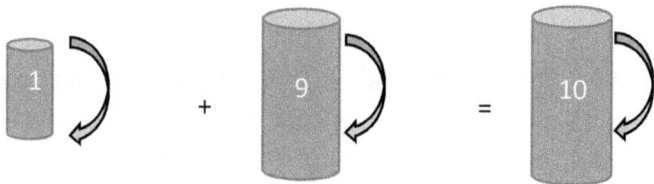

Yet, the same delivered package (quantum of energy) provides more energy if delivered at a faster rotation. But, it is very important to notice that it will only deliver and synch to a particle that has that same extra rotational energy.

As second example, with higher rotational energy, would end with a system with the same ratio of combined higher rotational energy.

So, if the particles do not match, they simply pass each other, there is not absorption.

Most use of Newton have been that the movement center remains constant. However, if you take that to the particle level, then the rotation of the system particles inside must also remain constant in the interaction. This is a nuance that is the key to the AVSC quantum-level model.

If Not in Synch, Approach Direction Same as Egress Direction

Let's do this for a zero Kelvin atom with an approaching photon.

Remember that there are multiple particles in this atom. Some will push and pull that approaching photon, but in the end:

> Total Momentum of System Remains the Same
>
> Total Rotational Energy of the system Remains Same

Whatever the photon pushes the atom approaching it pulls the atom at egress. Since they do not synch, there is a point where both

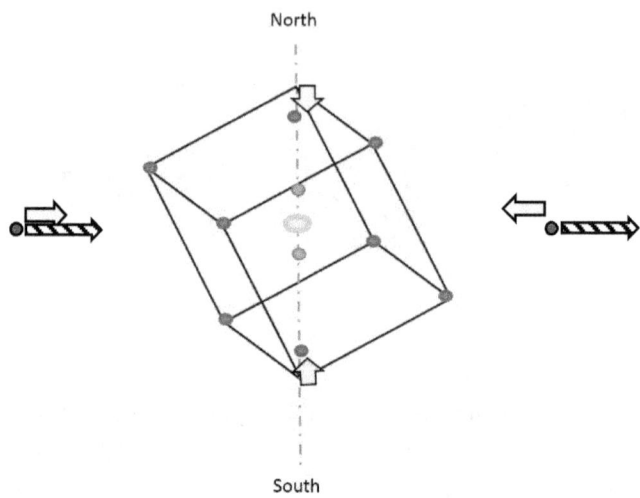

Even if it moves one of the atom's particles, it move it back – **_so long as it does not move it an entire position._**

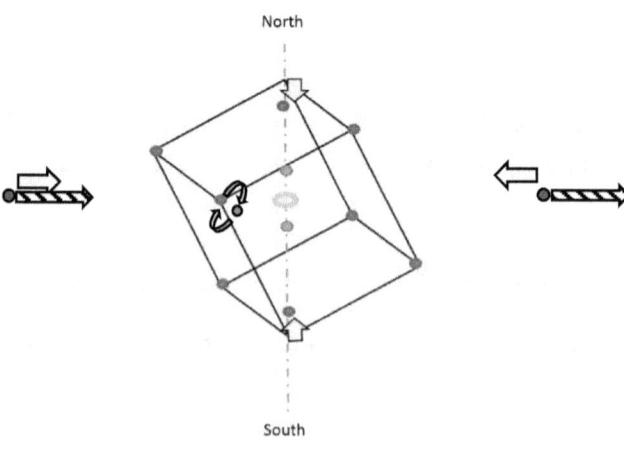

One step further: Nucleomagnetics Repulsions Moves Particle into New Settling Position

If the photon (purple sphere) and an atom particle (red sphere) has similar nucleomagnetics rotation (both beige arrows are at the same rotation frequency), then the atom's particle might combine and change the physical subshell position (black) of the particle. In the below example, the particle will move from 2c3 settling position to a different settling location 3f6 which has a) different energy (distance from nucleus and angle in nucleomagnetics field). That further leaves a Newtonian extra of a) overall momentum and b) overall rotational energy, which must be the ejected photon in a new direction with a new, different (see insert) frequency.

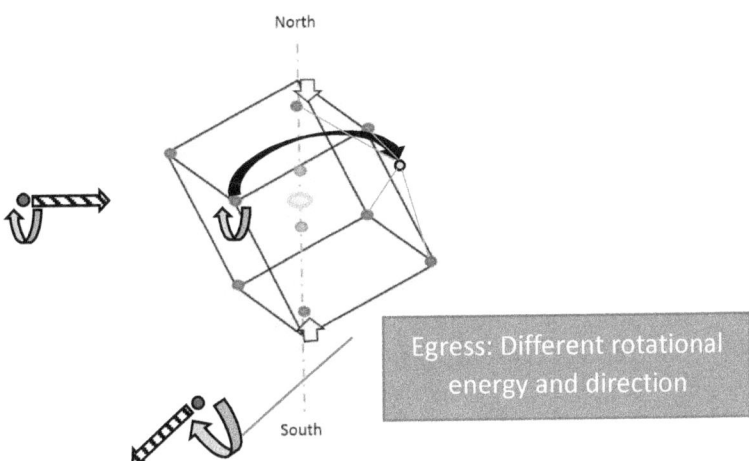

That absorption process is really that the photon can get close enough without the rotation creating nucleomagnetics repulsion.

In two examples, the two are at different rotations frequency, so they repel each other.

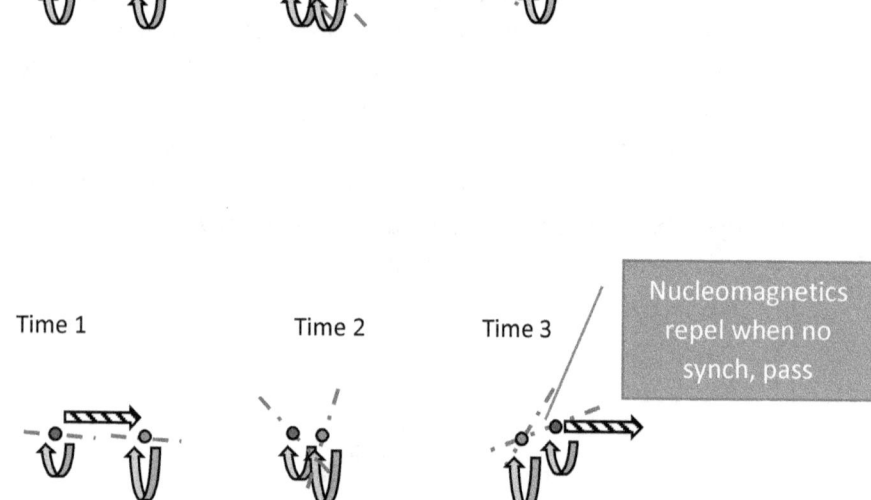

The mathematics of this is quite fun. This repulsion is not the straight electrostatic, motomagnetics is the change in the change of the rotating field strength. If not in synch, then the nucleomagnetics changes are relatively different – and the absolute value, so repelling. Repelling through the different rotational cycles.

Yet, if the two are in synch, then the magnetic strength change in each as it rotates remains also the same. That means at time, even with high rotation are the nucleomagnetics field strength different. That means that same frequency photons and particles have no time or position on the approach where the nucleomagnetics repulsion applies. Hence, they can get close enough to combine and create momentum to move the particle to another settling position (if the direction of the photon puts momentum towards that position.

More interesting, if the photon is absorbed, and its direction does not change the particle (electron) subshell settling position, then the photon must eject. Not every photon of matching frequency creates an electron re-positioning. That is why we get back to statistical methods and quantum mechanics.

I expect a number of PhD's to get granted exploring this issue more deeply. What I have covered is only a surface analysis extrapolating the AVSC Atomic Model to a limited level. My passions and time allocation are much more towards the periodic chart, and chemical reactions, so I will likely leave this to others.

Where AVSC Fits into Physics Theory?

Current theorists think of gravity (and mass) as an extension from Relativity. (170)

It projects that the solution for gravity is going further into relativity as a combination of General Relativity (from the top) and Quantum Field Theory (from the left).

As such, it requires curved space/time.

The Arno Vigen Scrunched Cube Atomic Model moves in the opposite direction.

1) Newton gravity is a simplification for basic particles of protons, electrons, and neutrons (not new subatomic particles). The AVSC model fits between Classical Mechanics and Newtonian gravity.

2) Further, in AVSC, Electromagnetism has been split into two Newtonian concepts for the two very different fields involved.

(171)

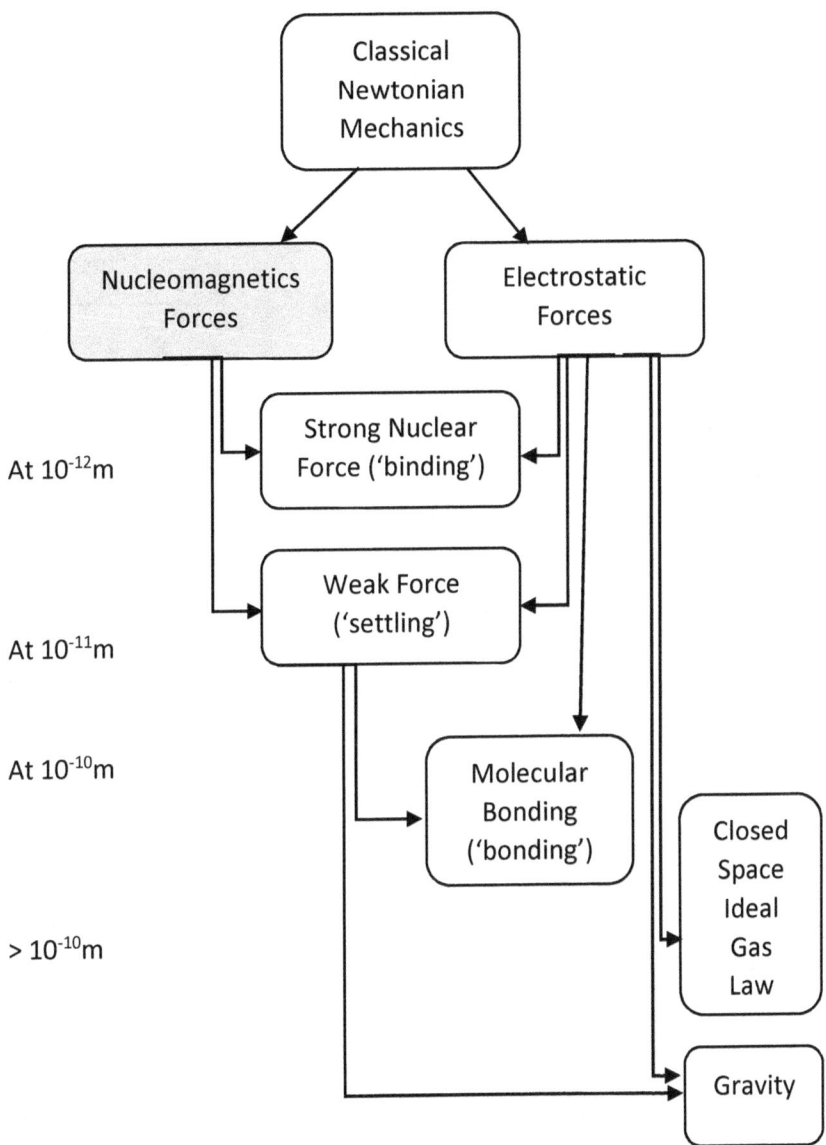

(172)

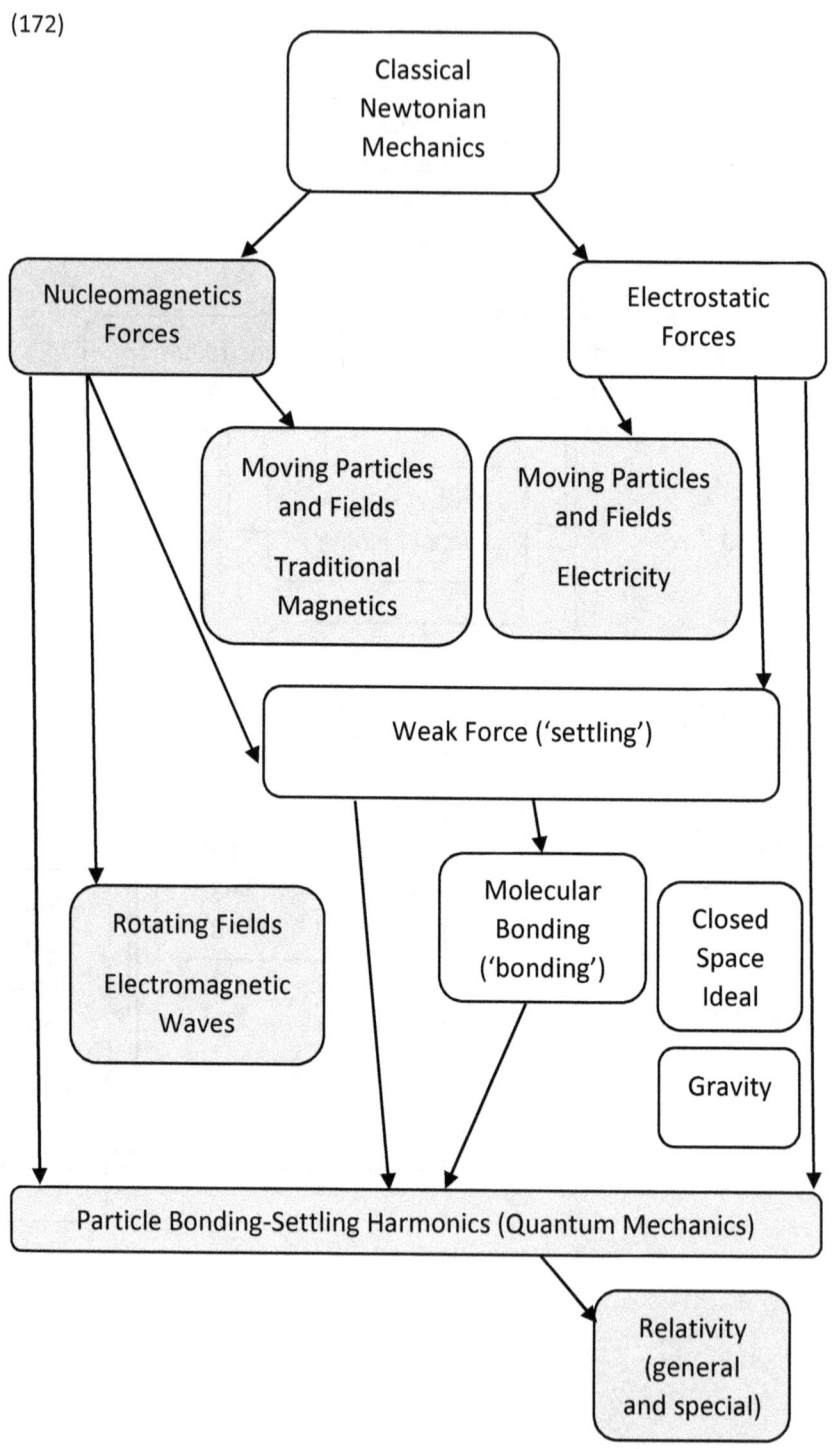

Therefore, in AVSC, Relativity is a tail, not a cornerstone. The cornerstones of AVSC are Newtonian opposites of:

- Electrostatic Charge
- Nucleomagnetics

Each derivative force (strong/binding, weak/settling, states-of-matter/closed space ideal gas law, molecular bonding, gravity) comes from those two **based upon the distances and particles involved**. That is the first chart in white boxes.

Each further observed force (traditional magnetism, electricity, electromagnetic spectrum, quantum mechanics, Lorenz, and relativity) comes from those two fundamental forces **based upon the rotation and movement**. That is the second chart in beige.

The nuances of a magnetic field versus an electrostatic field are not covered fully here, and will take another volume to discuss. However, as a teaser, think of a bedsheet grabbing one part to hold up as electrostatic charge, and the air holding it up as the magnetic field. That means when you pull at point up, you get this 1/distance-squared, but when you let go, the field does not stay in that shape. It stays up, and takes a different shape based upon 'trying to stay where it was.

The Arno Vigen Scrunched Cube (AVSC) Atomic Model is all continuous and deterministic. The addition of nucleomagnetics makes the current indeterminate Schrodinger's equation resolve as convergent in AVSC when the force on each particle includes both electrostatic and nucleomagnetics. The imaginary number 'i' is replaced by the nucleomagnetics force calculation such that we never have that $\sqrt{(-1)}$ ('i'). The missing dimension is the nucleomagnetics force. That proof also I will leave to another volume.

Endnotes

---

[i] By the way, the $= \dfrac{1}{\sqrt{1-\frac{v^2}{c^2}}}$ Lorentz transformation falls apart for molecules, let alone solid objects getting anywhere near the speed of light. That formula is only an approximation valid for v<(1/4)*c.

[ii] For reference, the magnetic field gets weaker at the ends and stringing around the equator (think 'bagel'). Therefore, the calculation of that requires an orientation strength.

In the larger sense, you can understand that beyond Hydrogen, the electrons are in all directions, and as such, those orientation ups and downs even out. The calculation is that amount of energy enclosed. That calculation is much easier.

Energy of N-Electrons repulsing each other = Energy of M(P+N) surrounded magnetic field.

$$N(E) \dfrac{kQ}{\left(\dfrac{r}{n}\right)^2} = N(P,N)\left(1 + \dfrac{r}{n}\right)^{nt}$$

Taking the separation of electrons from each other, then the volume that separation creates (a sphere) must cover the area of the magnetic field. This calculation can get down without the 'bagel' complexity.

Electrons push each other away, but protons pull them together. Protons push each other away, but electron get attracted to them. All those particles have the same Charge. To each other, once you cover more than

This calculation is similar to a famous work on Bohr radius a hundred years ago.

[iii] These parts can get calculated easily.

The first shell has little electron-electron repulsion so it isolates the proton attraction well. Example of shell size balance can get seen in the Shell-1 (1p/1m) distances found in existing standards.

[iv] http://feynmanlectures.caltech.edu/I_01.html

[v] https://en.wikipedia.org/wiki/Chemical_bond

[vi] Using Coulomb's Law, the Charge-Force at 1-meter is:

$$F = k_e \frac{q_1 q_2}{d^2}$$

Factor	Net-Charge Calculation
Charge Force factor	$k = 10^{10}$ m² / (s²) [9.03×10⁺⁹]
Charge of orbiting Electron	$Q=10^{-19}$ [1.602×10⁻¹⁹]
Charge of distance Proton	$Q=10^{-19}$ [1.602×10⁻¹⁹]
Distance	d=1 m or 10^0
Exponent shortcut	+k+Q+Q-d-d
Gross Charge Short-cut calculation	10-19-19-0-0 =
	10-38-0 = -28
Gross Charge Force	10^{-28} m¹ / (s²)

[vii] https://en.wikipedia.org/wiki/Proton

[viii] http://www.bing.com/search?q=distance+of+Bohr+radius&src=IE-SearchBox&FORM=IENTTR&conversationid=

[ix] https://en.wikipedia.org/wiki/Quantum_gravity#/media/File:Quantum_gravity.svg from original by Raidr vectorisation by B. Jankuloski

www.ingramcontent.com/pod-product-compliance
Lightning Source LLC
Chambersburg PA
CBHW050216230526
45470CB00001B/404